COAL MINE

COAL MINE

Work in the Mine

Photographs and Texts
Friederike Most

EDITION STEMMLE

Contents

Acknowledgment

to all of the people who allowed me to photograph them, whom I would have liked to mention by name but cannot; and thanks for those moments of "refreshment;"
to Randolf, the Thirteenth Apostle, who proved that public relations and a friendly attitude are part and parcel, and who arranged a descent into the mine for me whenever I wished;
to my personal shift foreman Franz V., who constantly saw to my safety and made sure there was adequate lighting;
to Klaus "No-Problem" Vogel, the lord of the computer,
to Jörg Hildebrandt, Chef de Passepartout,
to Professor Detlef Orlopp, who taught me how to see,
and to Norbert Ulitzka, whose active enthusiasm for my work made so much surprisingly simple.

Work in the Mine

Friederike Most

Have you ever been in an underground mine? Have you ever wondered how it is down below, what it is like to work in a coal mine? You may be surprised to learn that today nearly 70,000 people are still at work mining hard coal underground (in Germany).

Over the centuries since coal mining began, work in the mines has changed considerably. Technological development has proceeded so rapidly during the past forty years that it is legitimate to speak of a revolution. What remains, however, is the miner, the human being underground and the virtually unchanged manual labor he must perform, labor that calls for total physical effort and demands a special inner commitment. People who see this as just a "job" would never be able to expose themselves day after day to the dangers of a miner's life. It is this inner commitment that has produced the strong sense of tradition still felt by miners today. Perhaps a sixteen-year-old novice miner would prefer not to admit to knowing the Steigerlied (a traditional miners' song), but a hearty "Glückauf," the salutation used by miners since the end of the 17th century,1 passes his lips as naturally as "hello."

While today's mining technology makes a number of tasks easier, it poses new dangers as well. Miners have always had to contend with hazards such as rock slides, mine-gas explosions and unforeseeable cave-ins. In our time, accidents caused by hydraulic hoses pumping an oil-and-water mixture at tremendous pressures, derailed or improperly maneuvered shuttles or mine cars, mechanical mining equipment (boring and milling

Translator's note: The significance of the expression goes beyond its literal meaning of "good luck," as it is an affirmation the rigors and dangers faced by all miners.

machines) and automated conveyors (armored shuttles and belts) must be reckoned with as well.

An essential precondition for work in the mine-shaft complex is the supply of fresh air, known as "weather management." Mine-shaft temperatures are regulated by means of a special method of calculating humidity, air temperature and air-flow speed. Anyone who has ever been down below in underground mine knows all too well what the expression "matte Wetter" (approx. "dull weather") means and is aware that the definition "bad air" doesn't convey the true sense of the term. Ordinarily, it is the temperature of the earth and the coal itself that heats the air. Introducing fresh, cooled air helps keep temperatures within acceptable limits. In the course of this year's collective bargaining negotiations, however, the work-climate regulations were changed, eliminating the local option to reduce maximum

daily work hours to seven under extreme temperature conditions. There is good reason to fear that efforts devoted to ensuring "good weather" in the mines may be drastically reduced.
Under the influence of the current situation in the coal industry, mine mergers and transfers of miners to other shaft complexes are splitting up work teams that have been together for years. As is true nearly everywhere else, miners' fears about losing their jobs and incomes simply increase the pressure that already exists.
This book is not intended as a complete and comprehensive appraisal, such as a history of coal mining might offer. It is meant instead as a memento – for those who work the mines today, for all those who have ever worked "down below" and not least of all for those who may be unable to imagine how deeply the lives of the people of the Ruhr region are anchored in that world.

While the conditions under which miners work may seem strange and far-removed to outsiders, the faces encountered here are familiar ones: people at work.

Accompany me on this brief journey of exploration into the life and work of miners "down below."

12

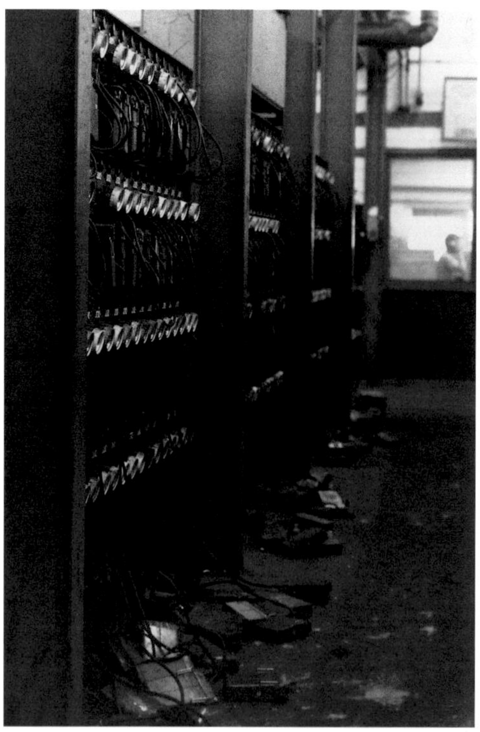

“Lampenstube”
(room where miners’ lamps are stored between shifts)

CABLE LIFT DESCENT, FLOOR 885 m
SHIFT CHANGE

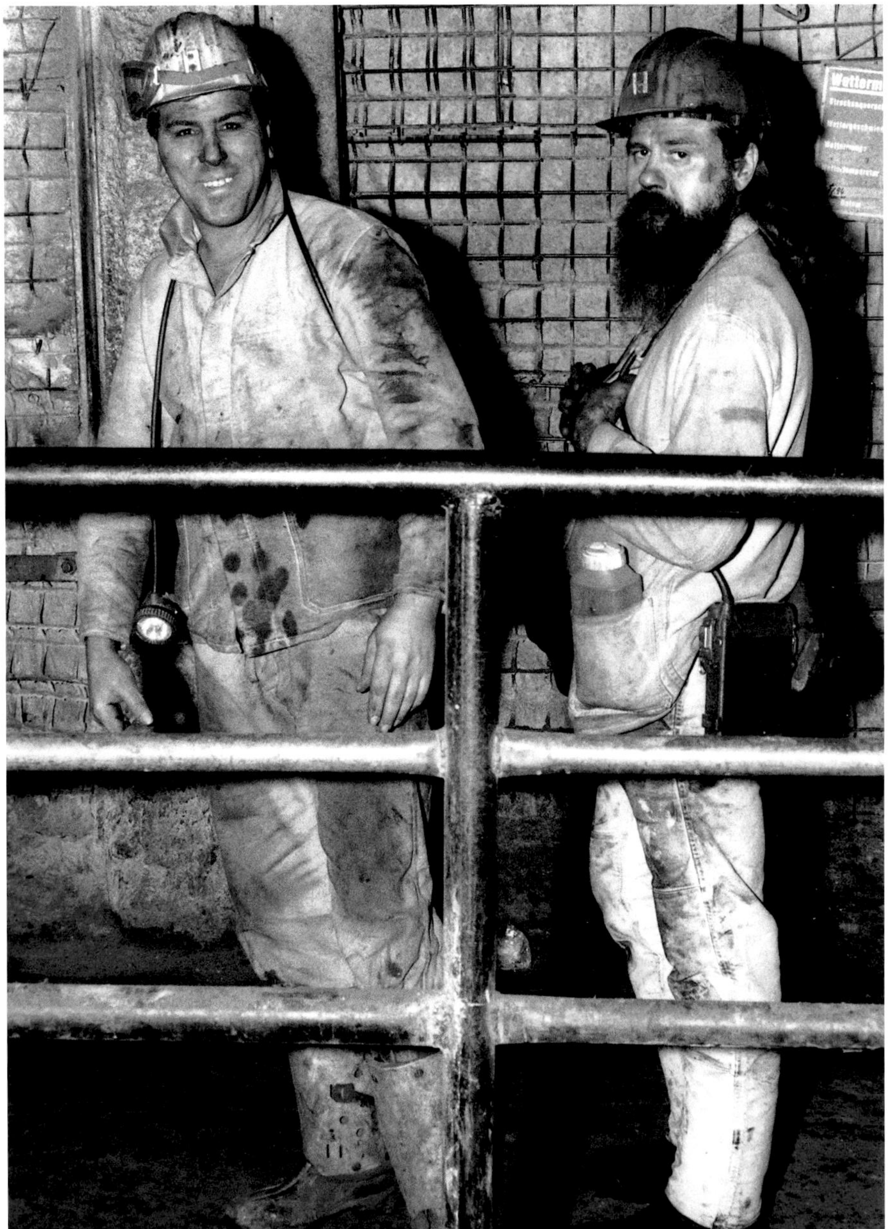

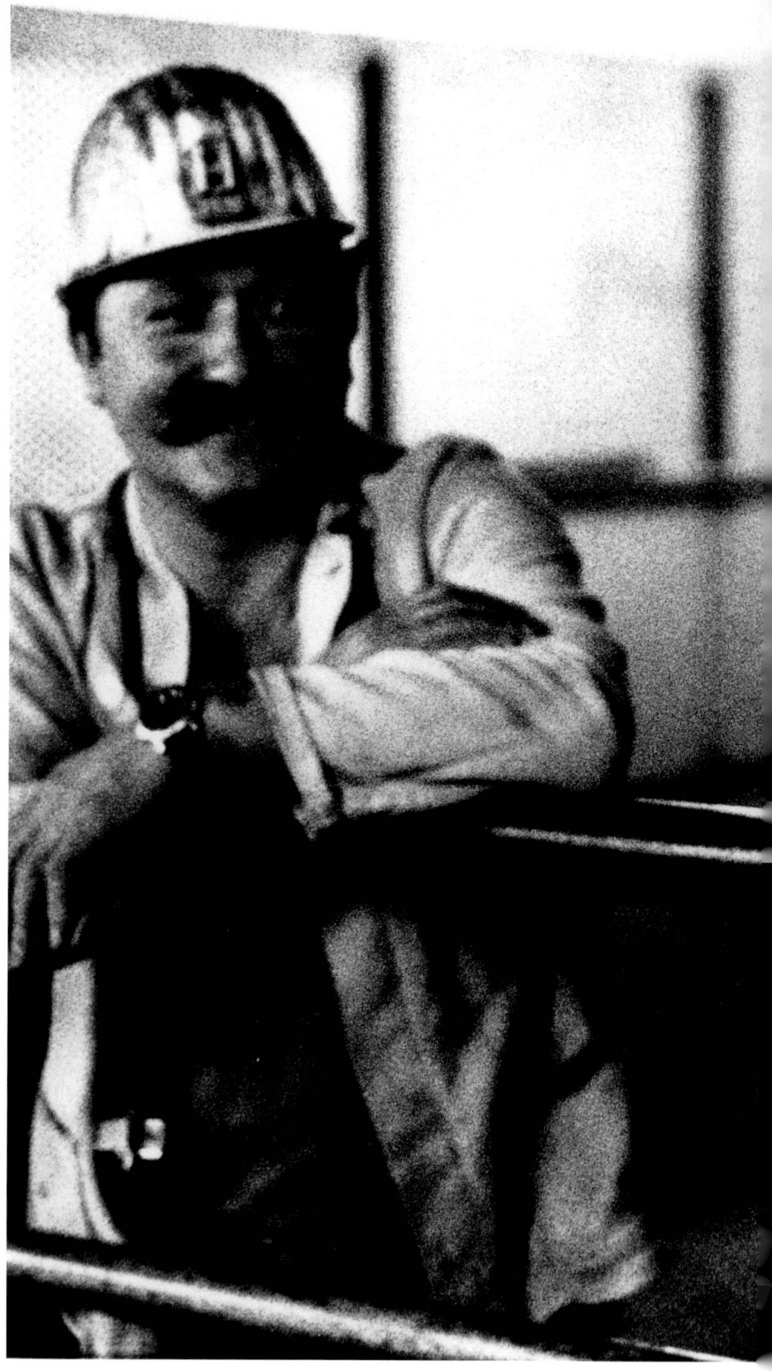

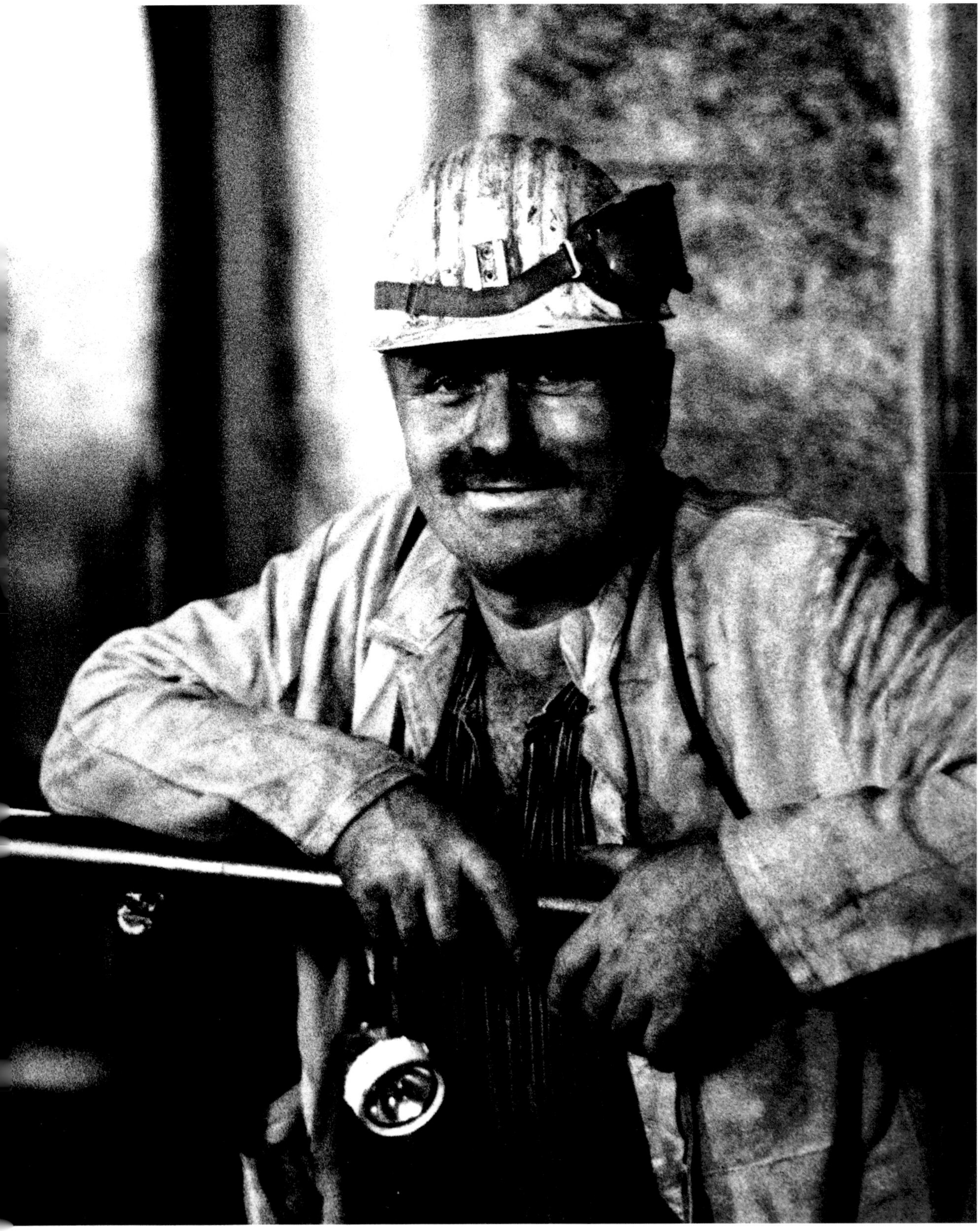

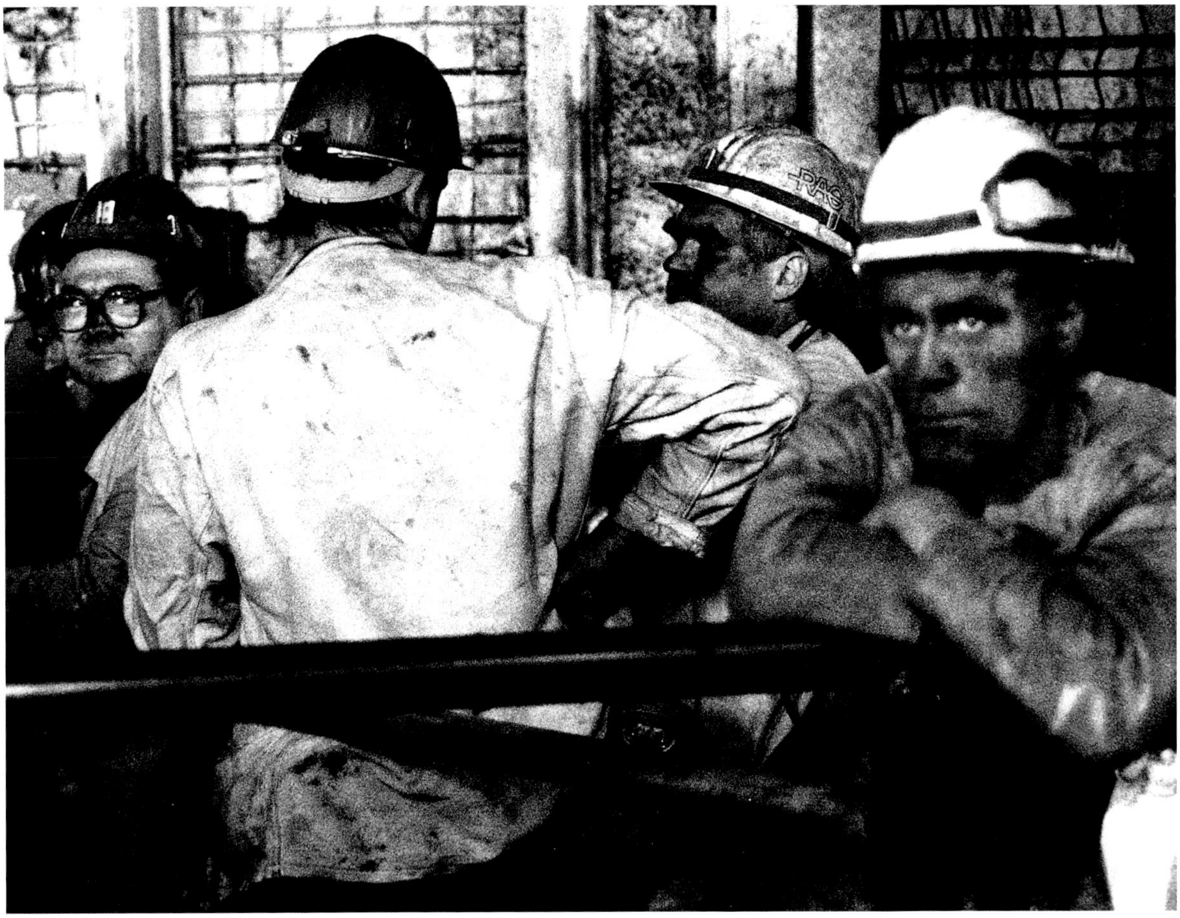

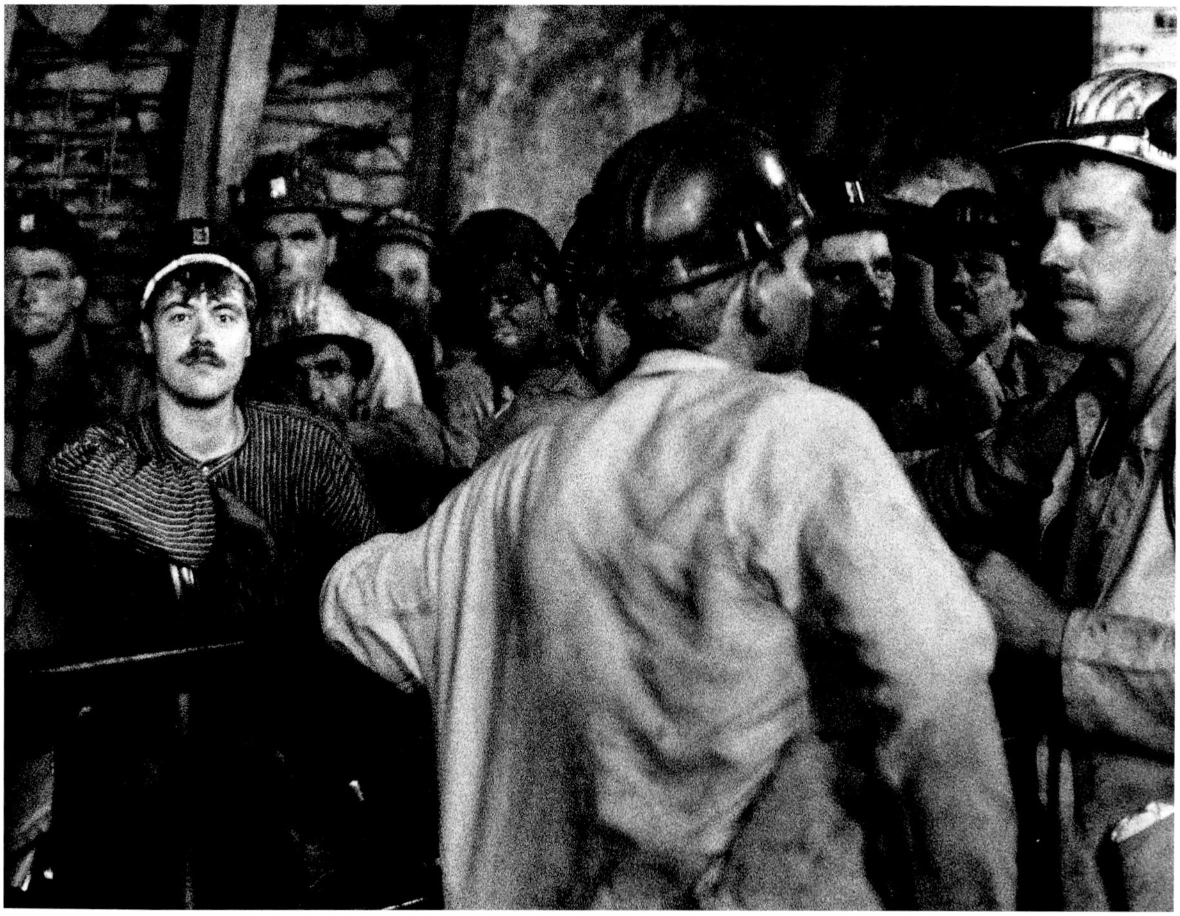

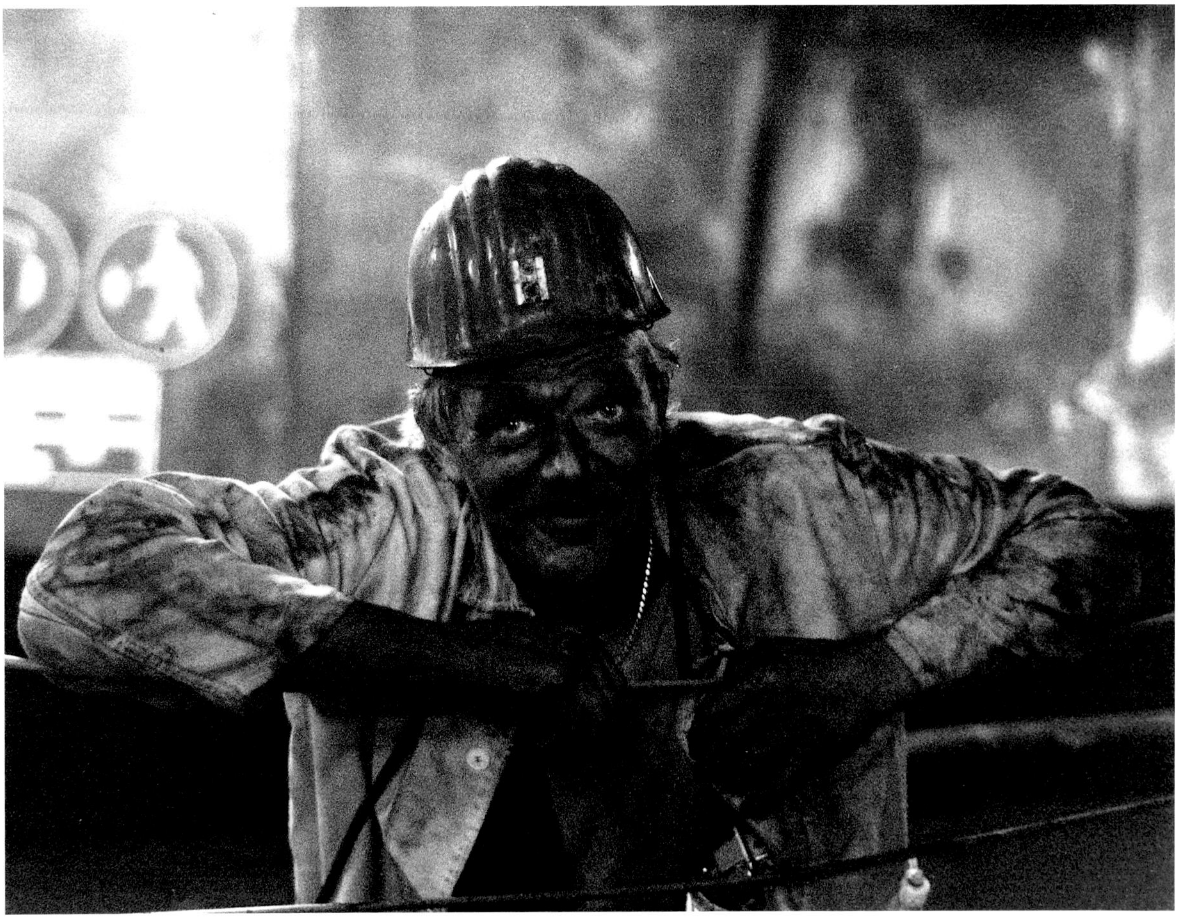

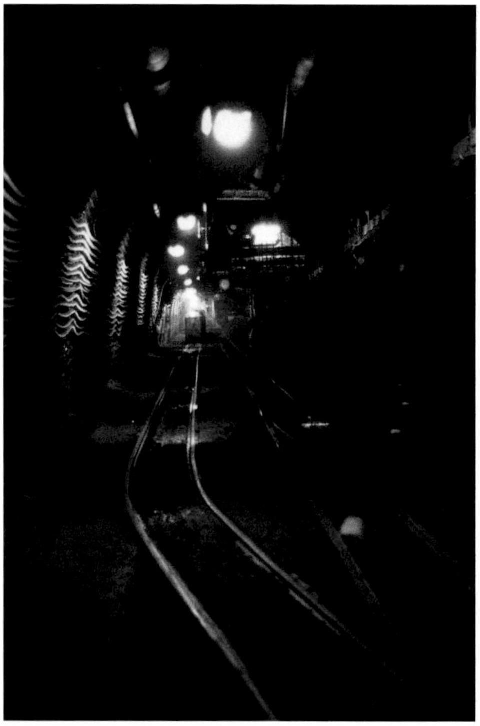

Horizontal shaft with rails

WHEN ALL WORK BEGINS

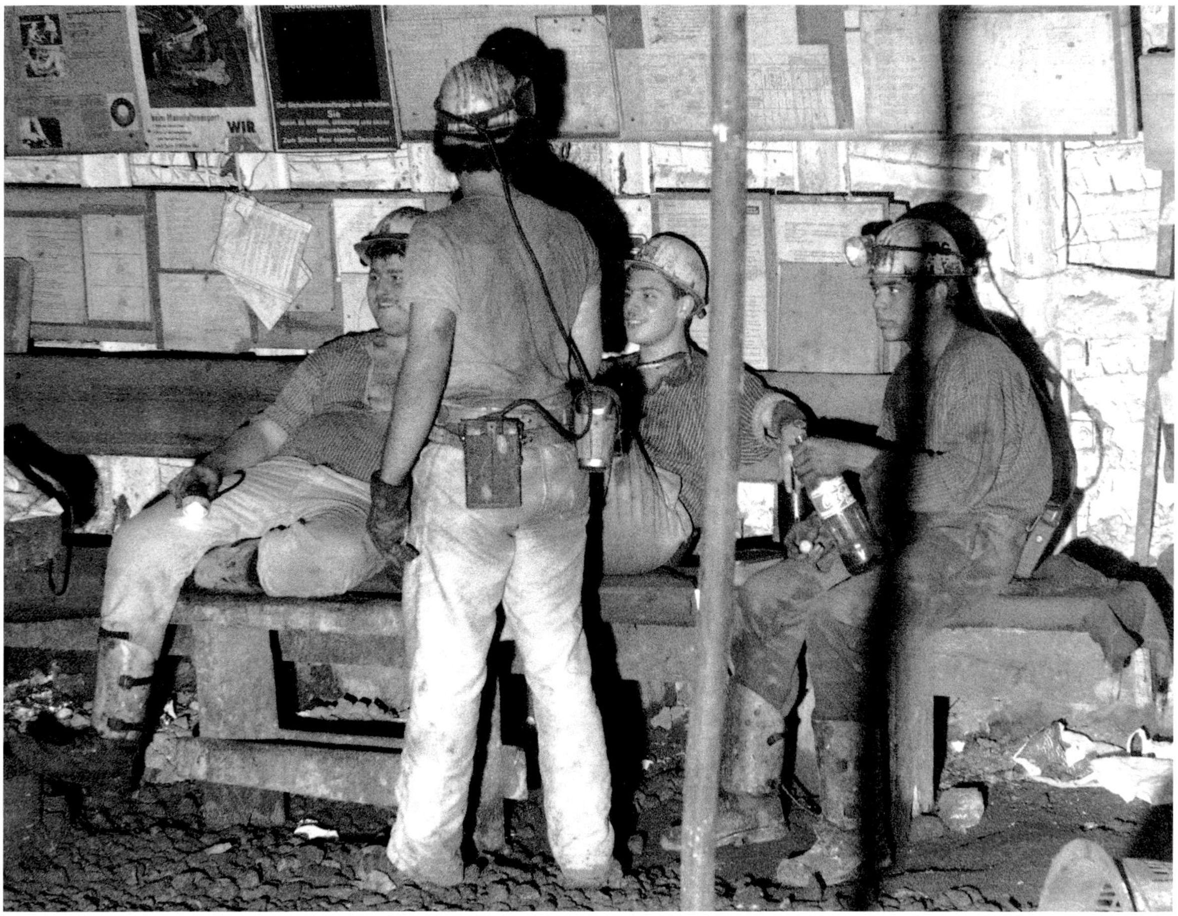
WIR
Sie

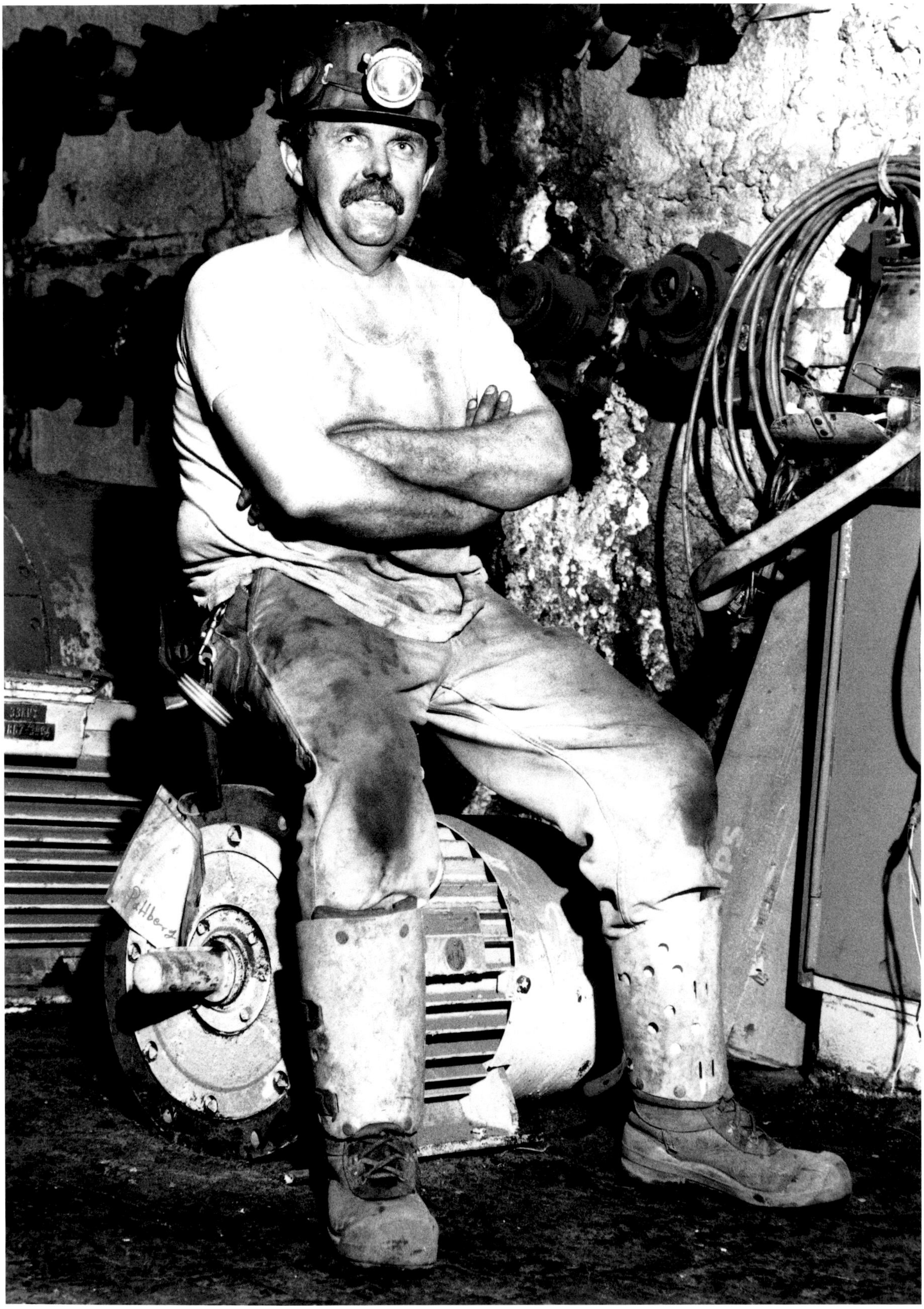

28

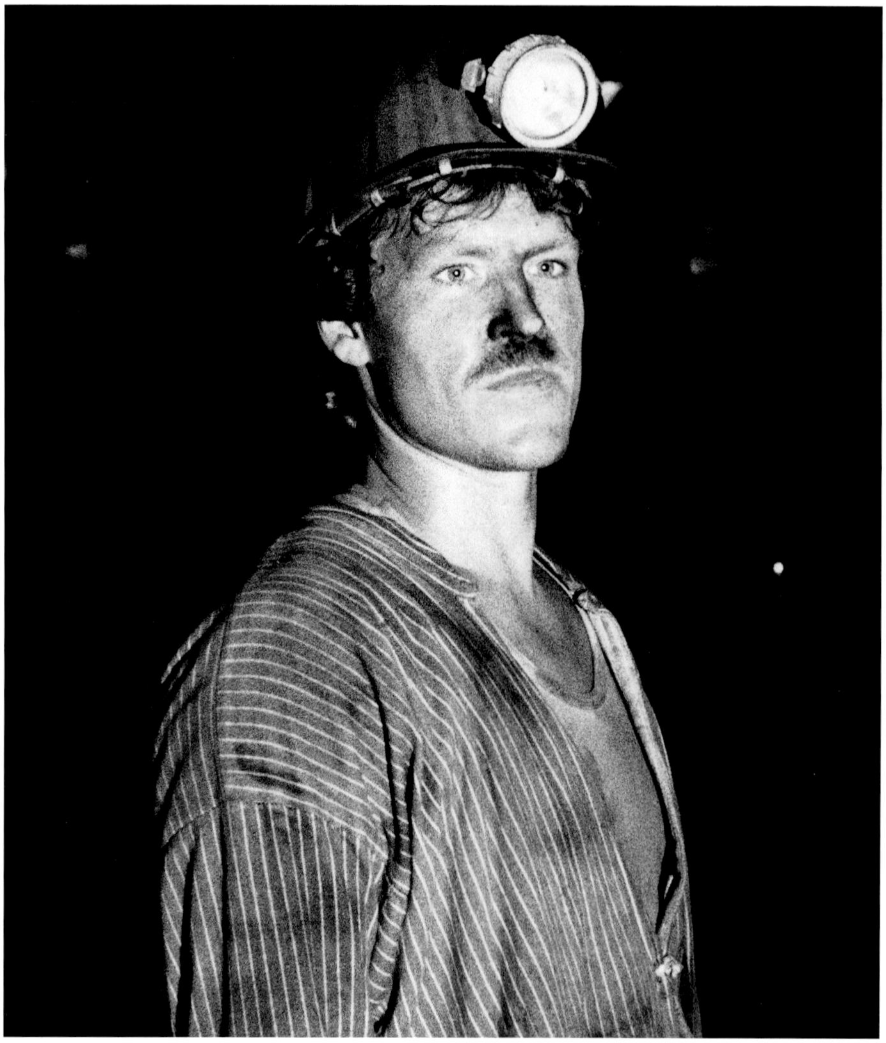

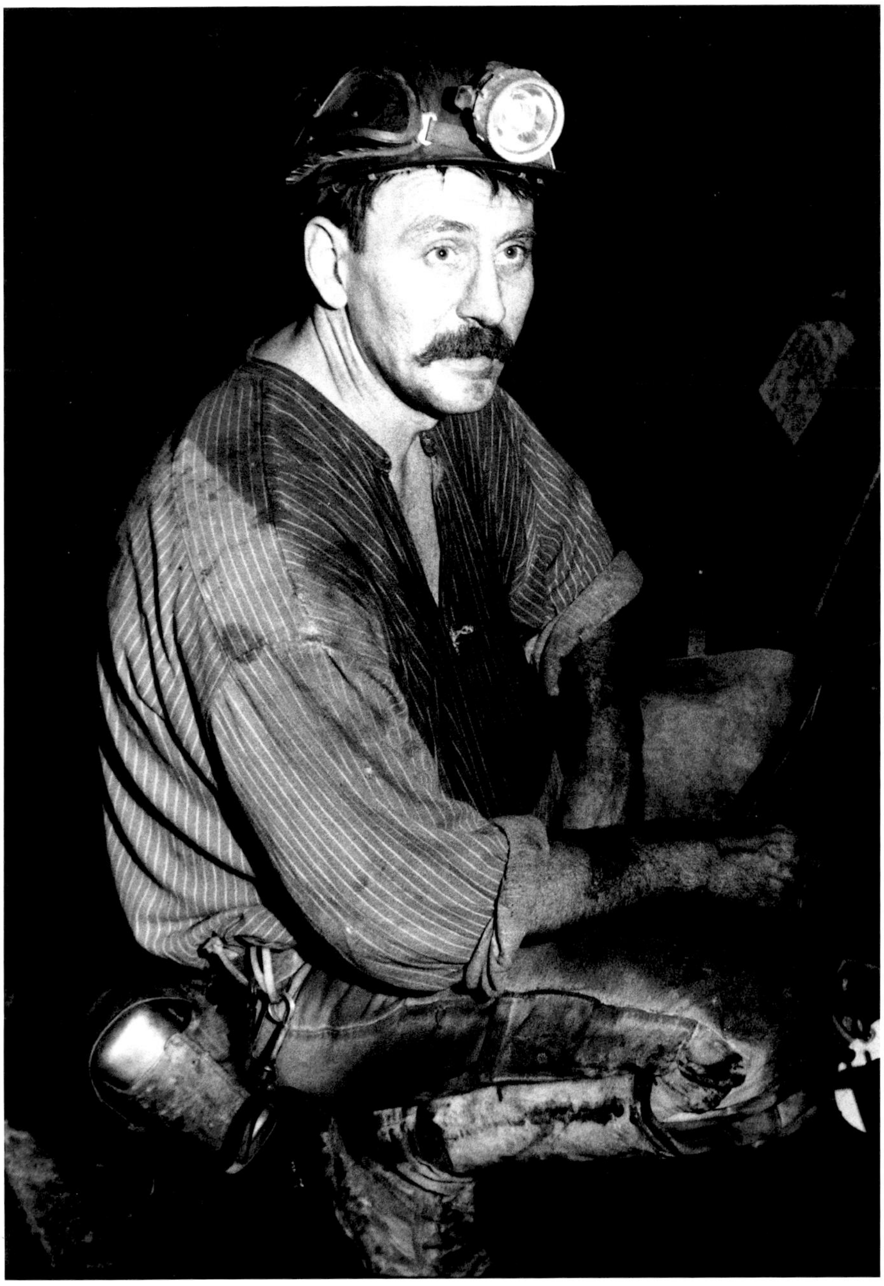

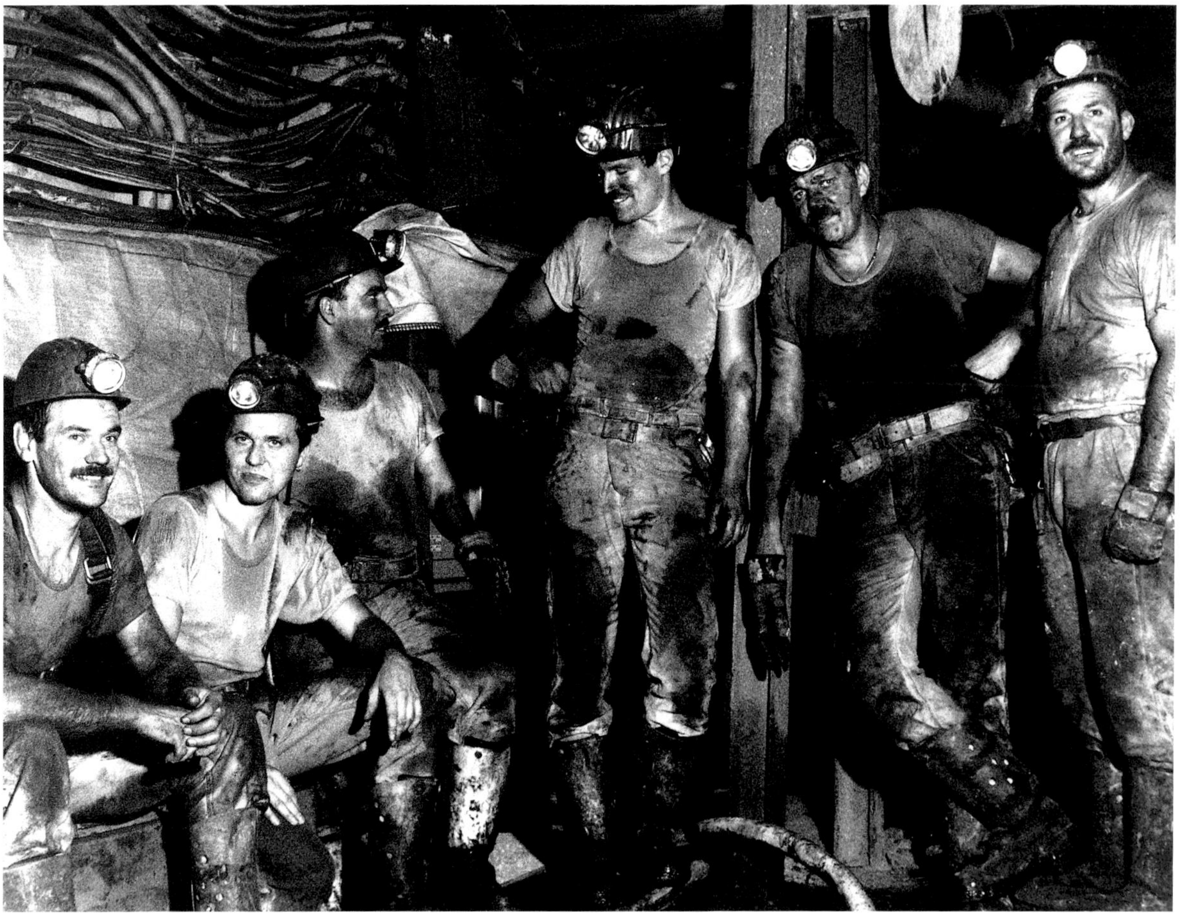

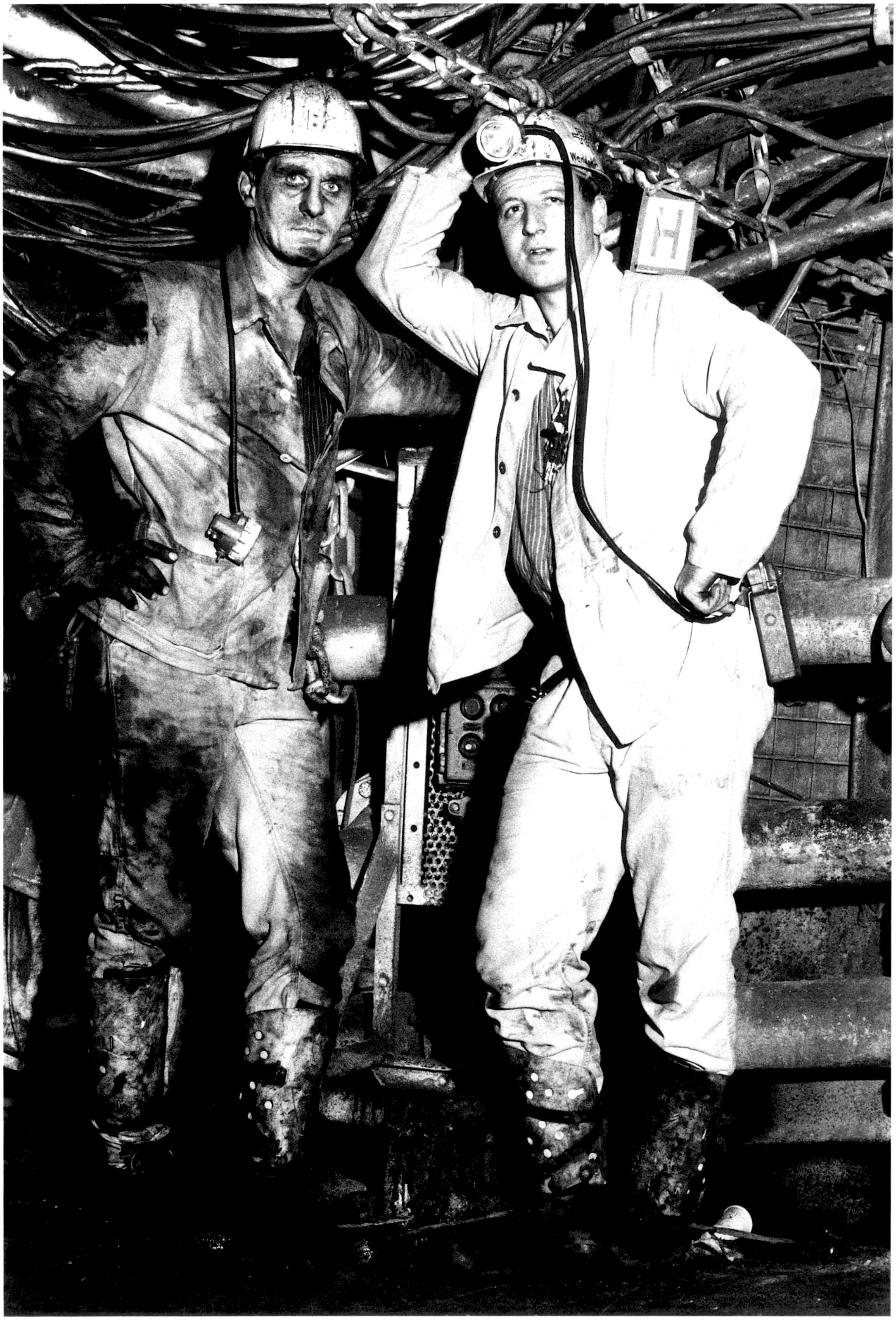

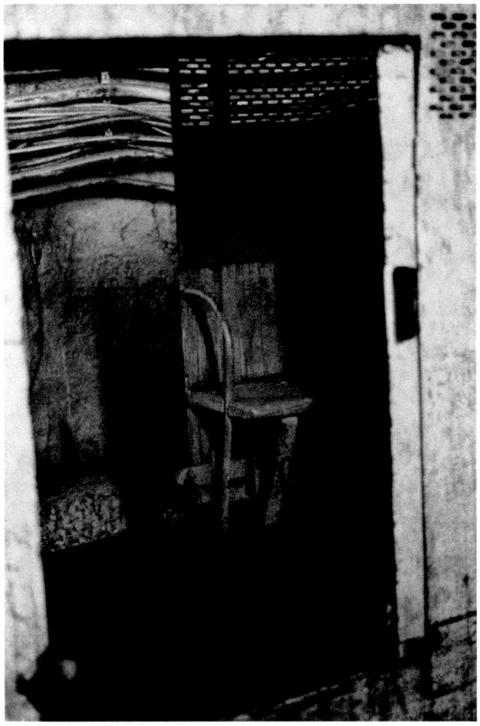

View inside the miners' cage

HOPPER RAIL LINE 03 55,
TUNNELING

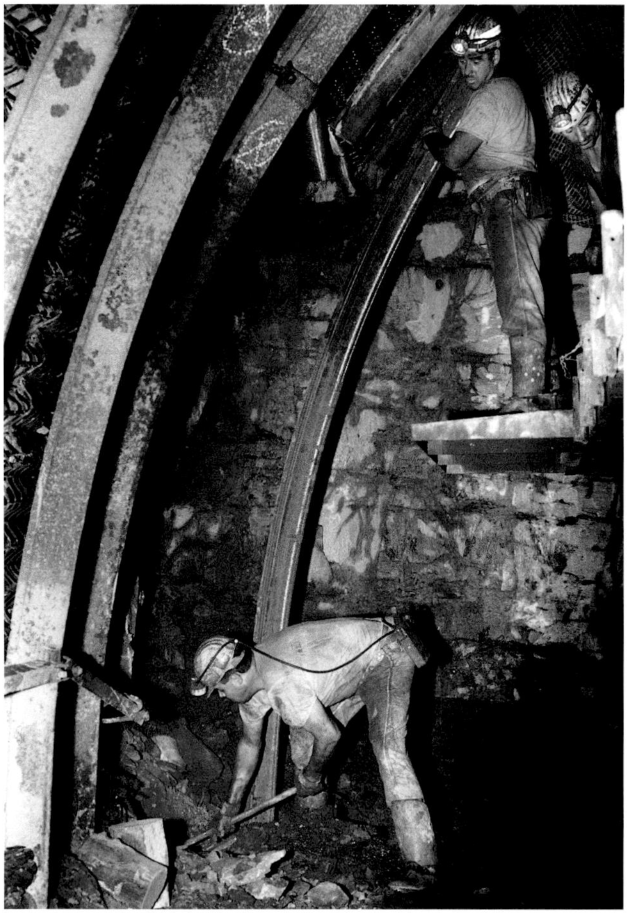
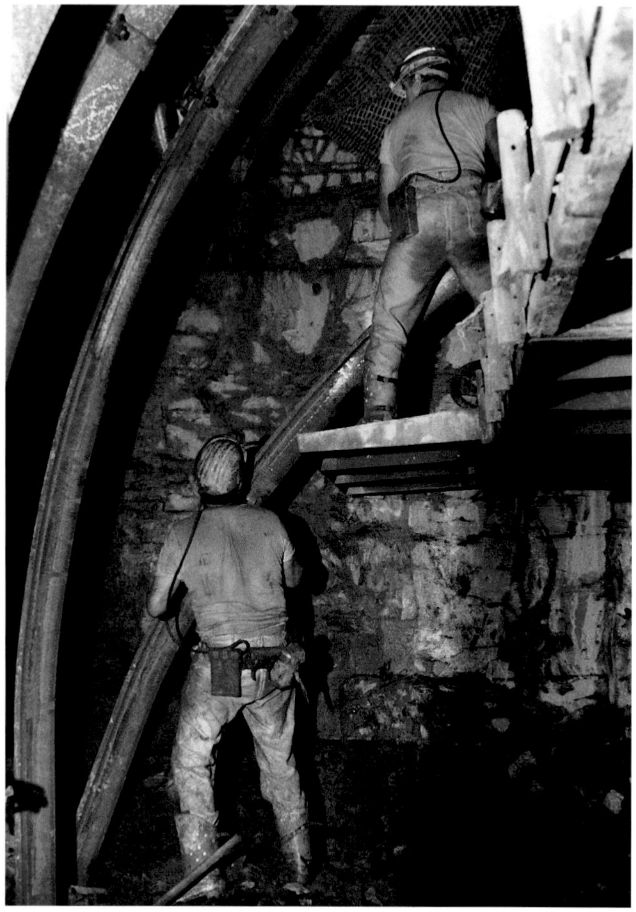

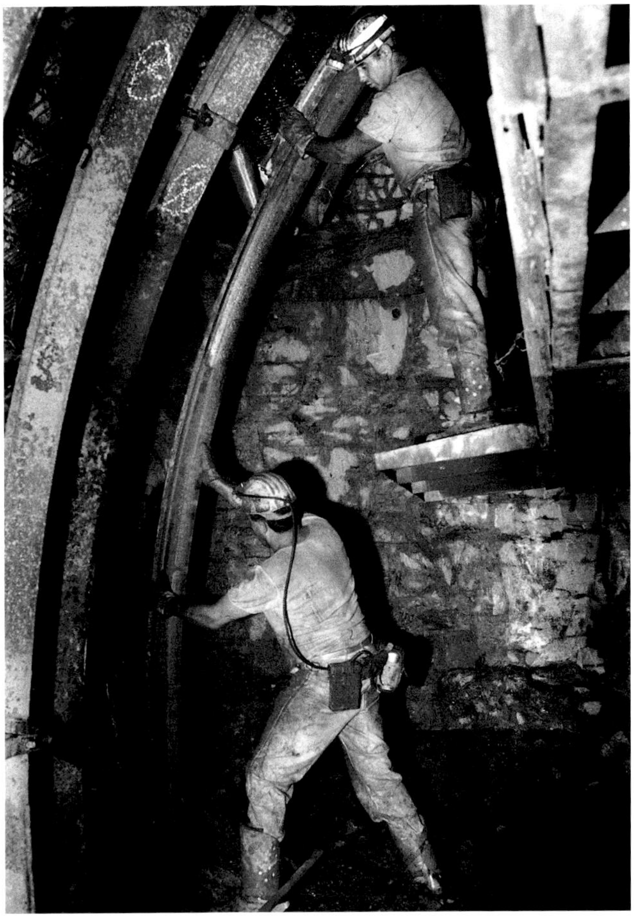

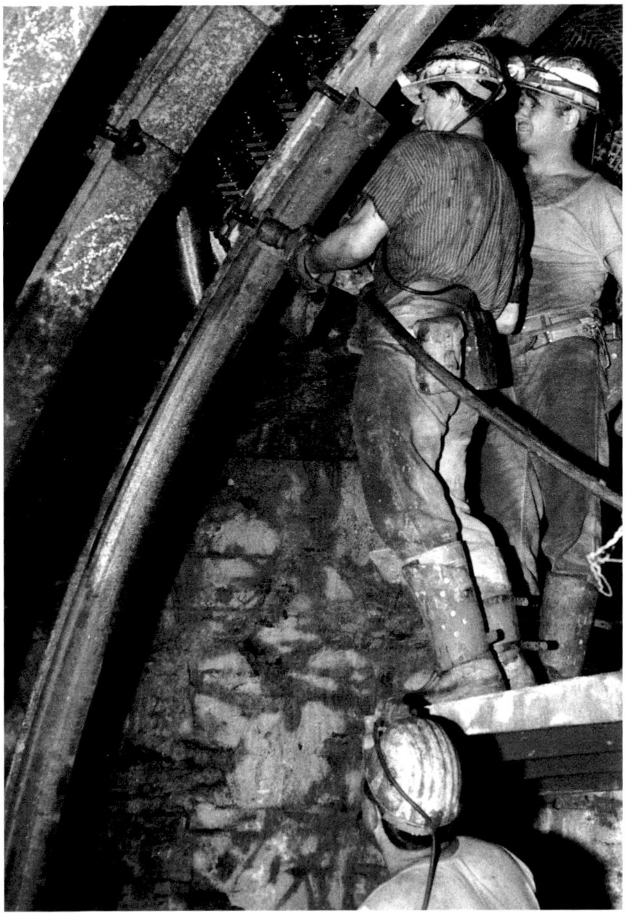

Durchschlag – a planned or unexpected breakthrough into an existing shaft complex while producing a new one.

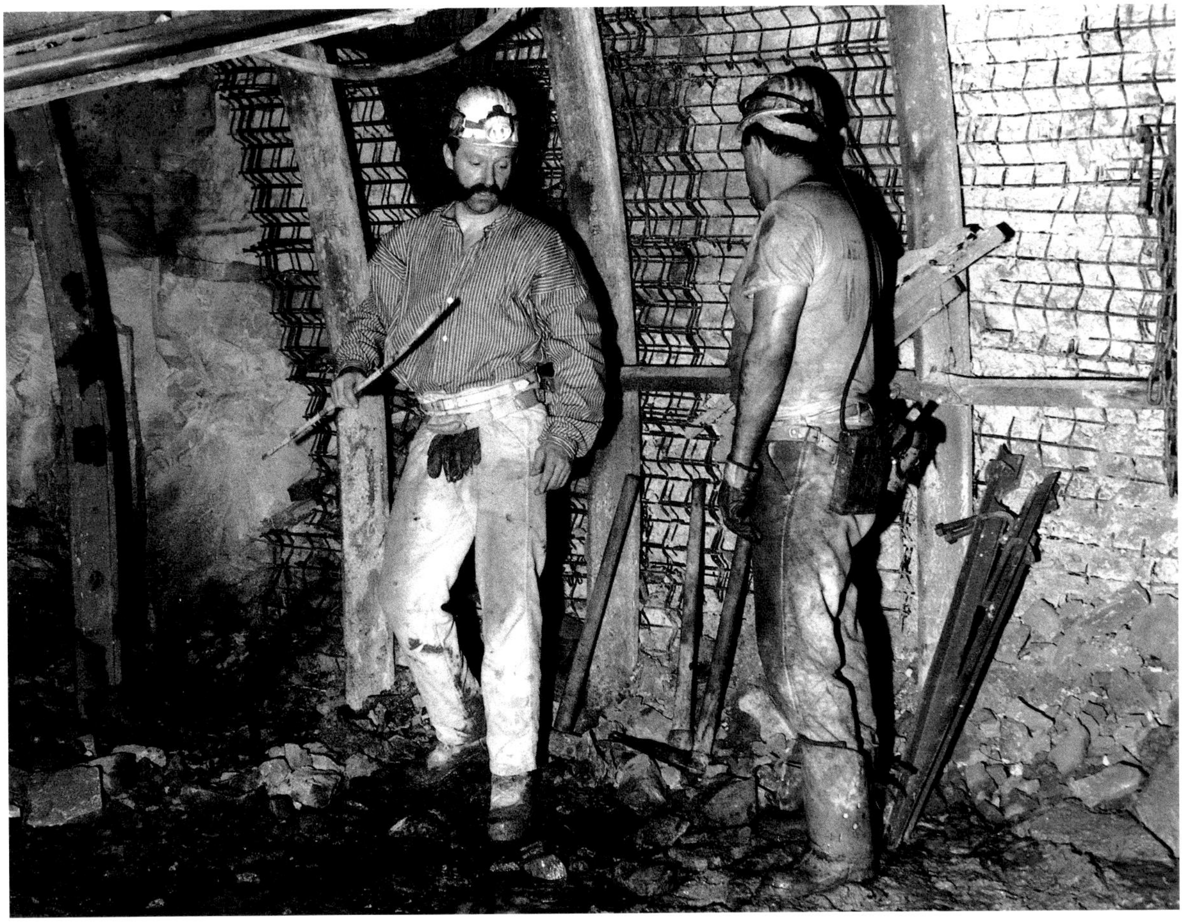

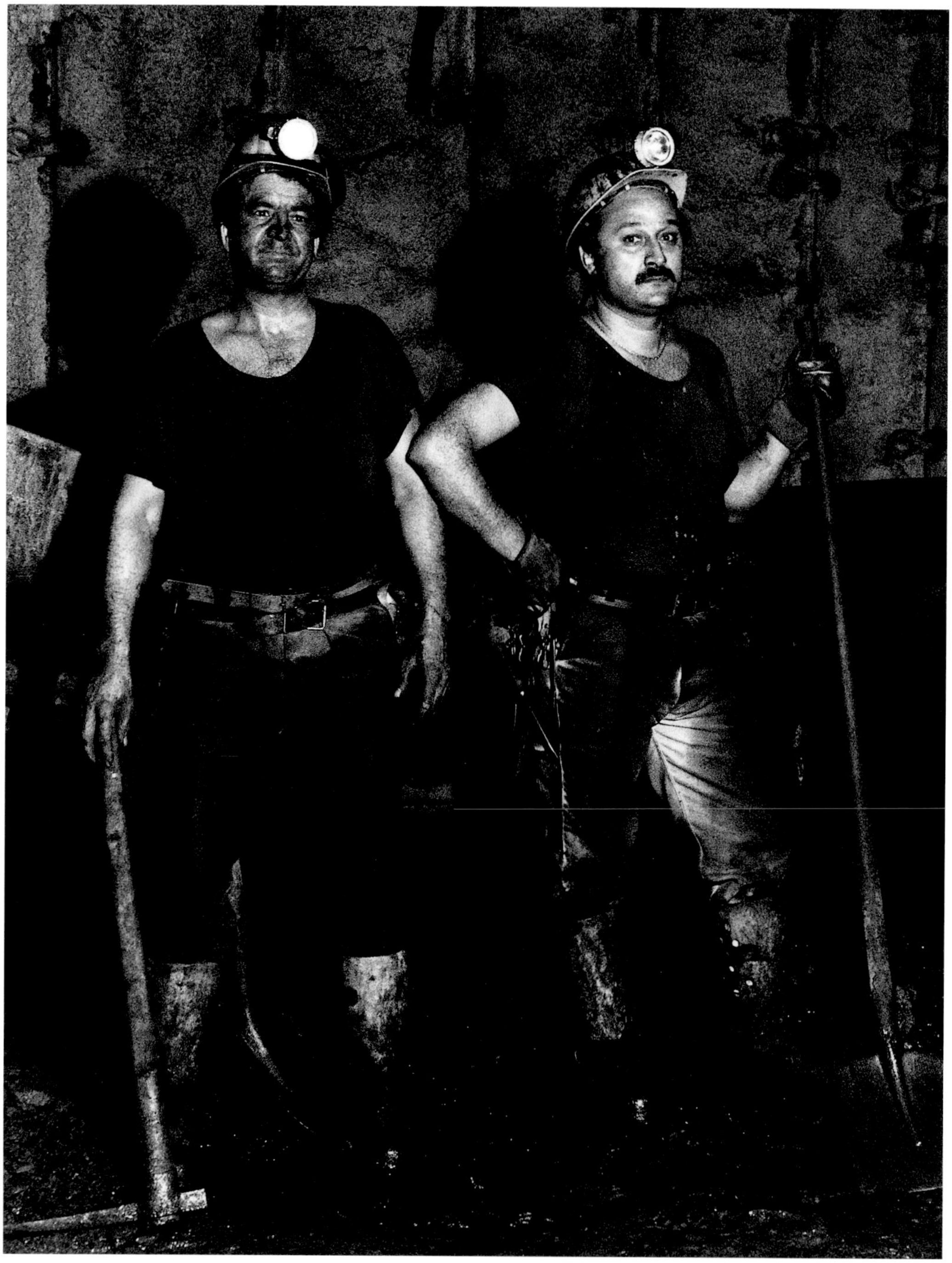

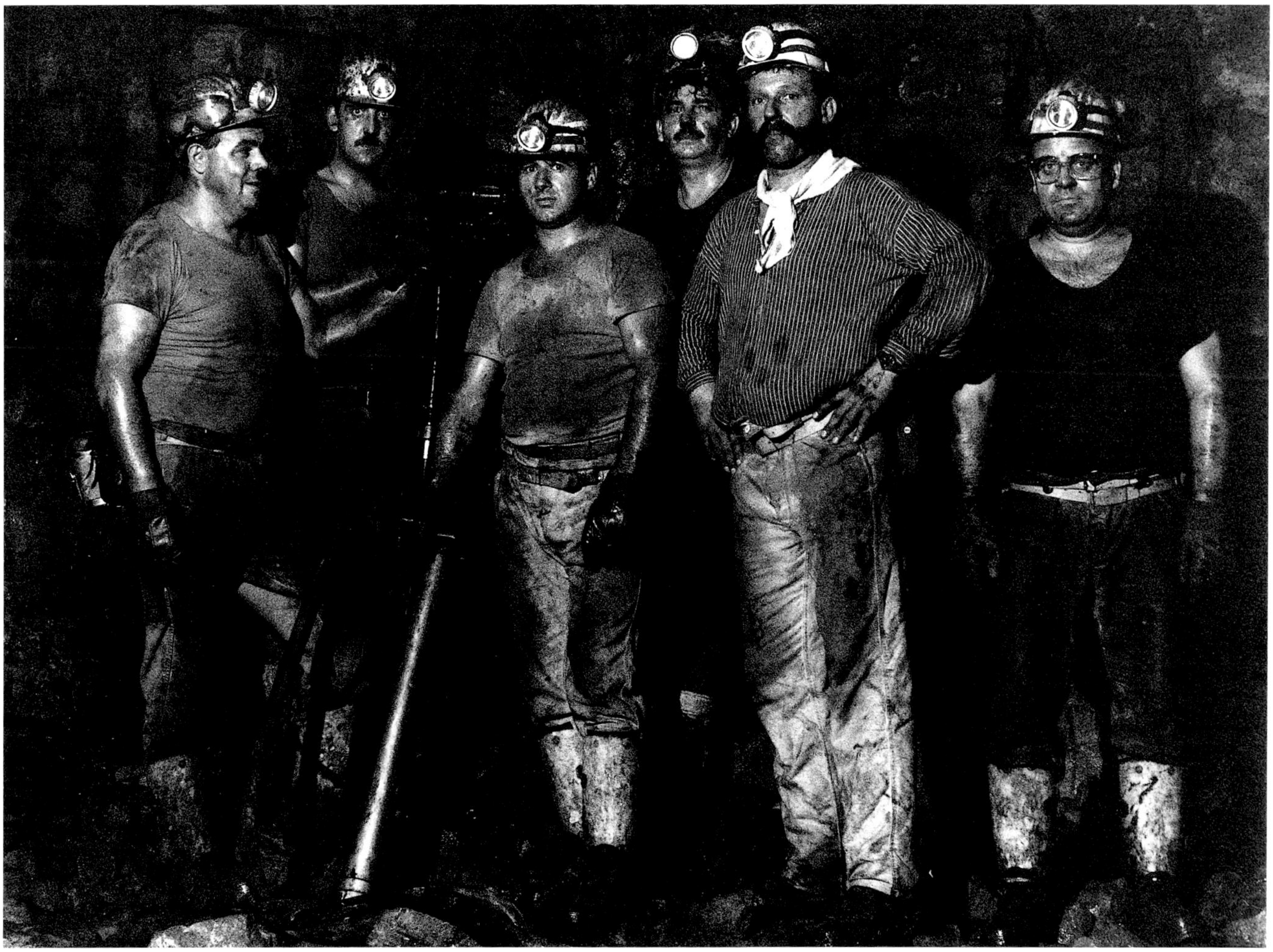

Travel time – the time required for movement underground from the main shaft to the workplace and back.

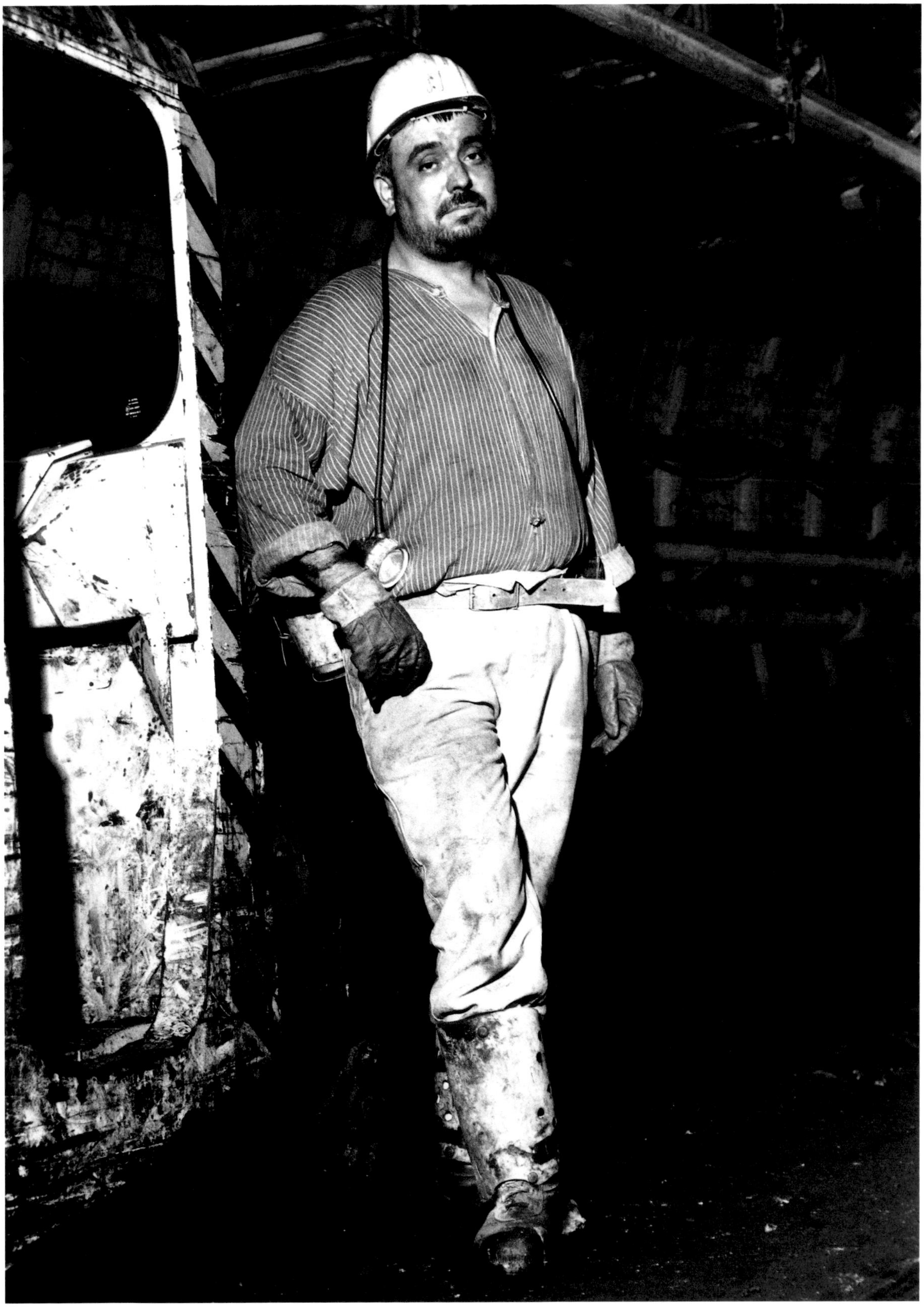

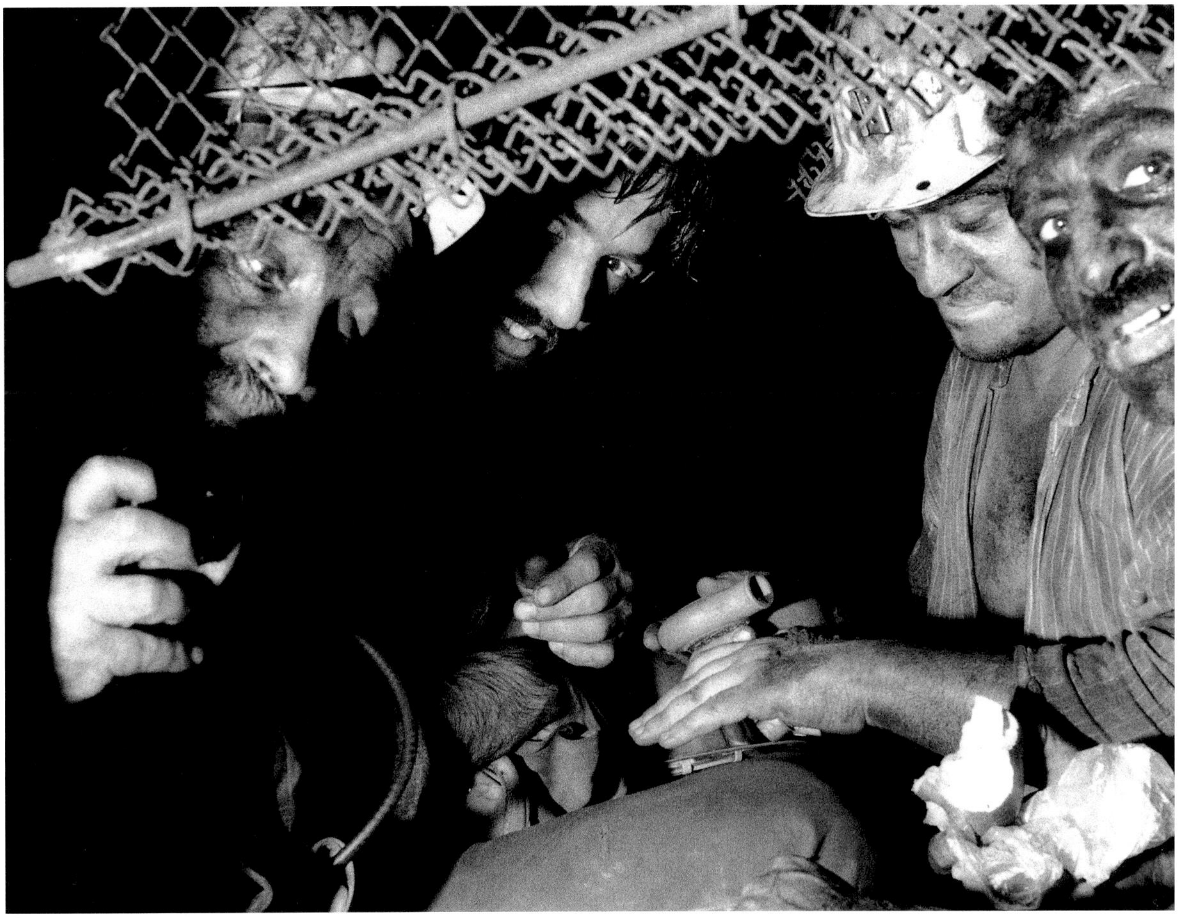

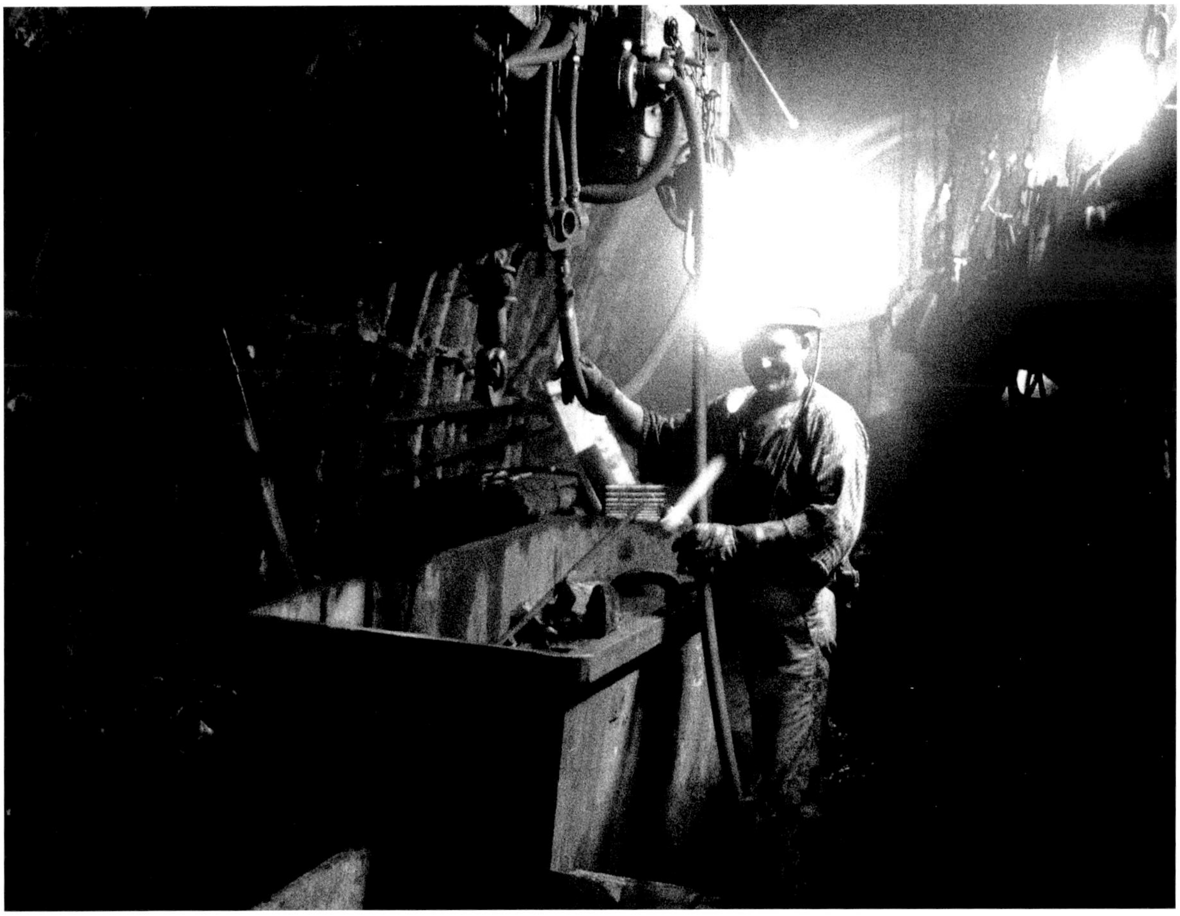

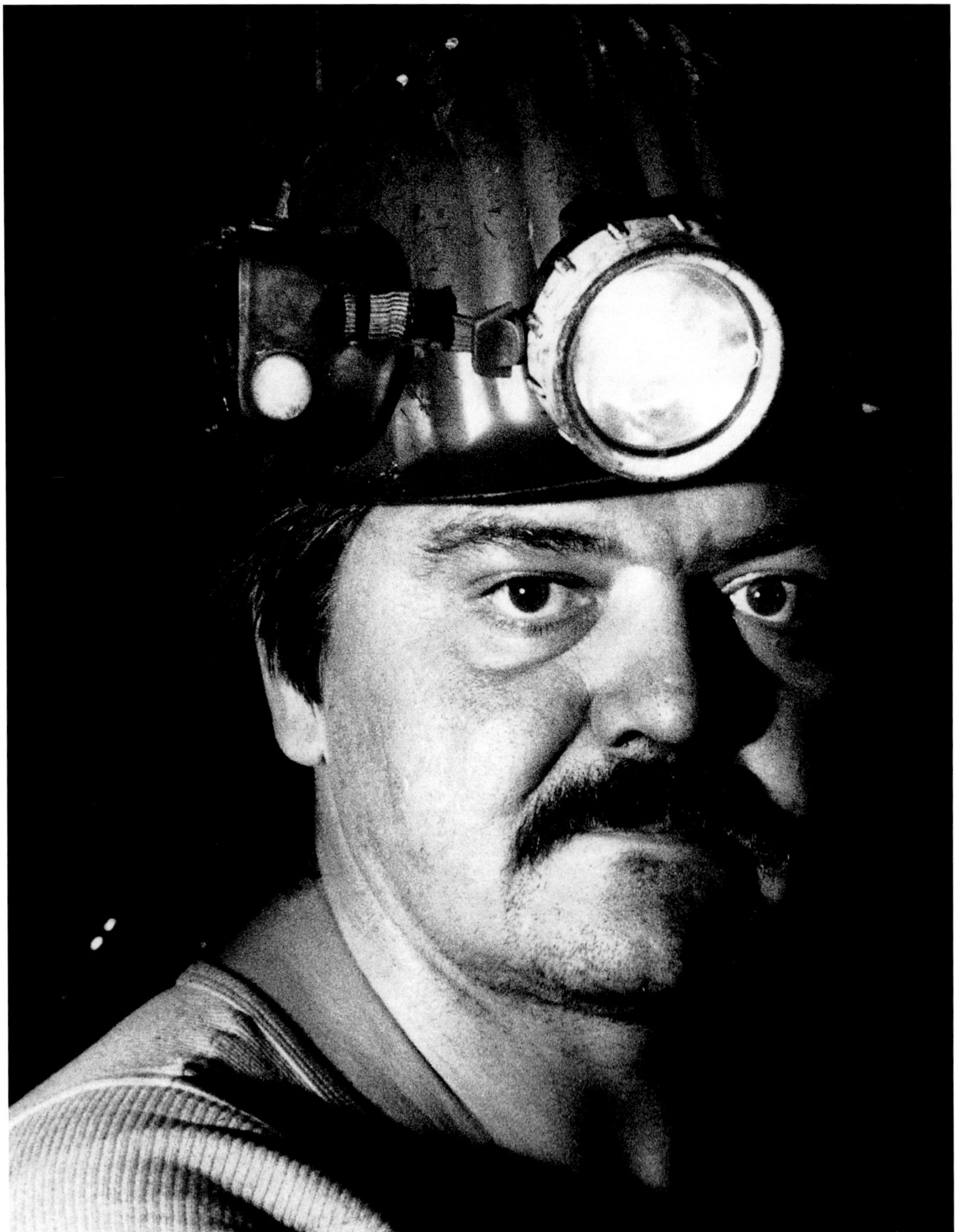

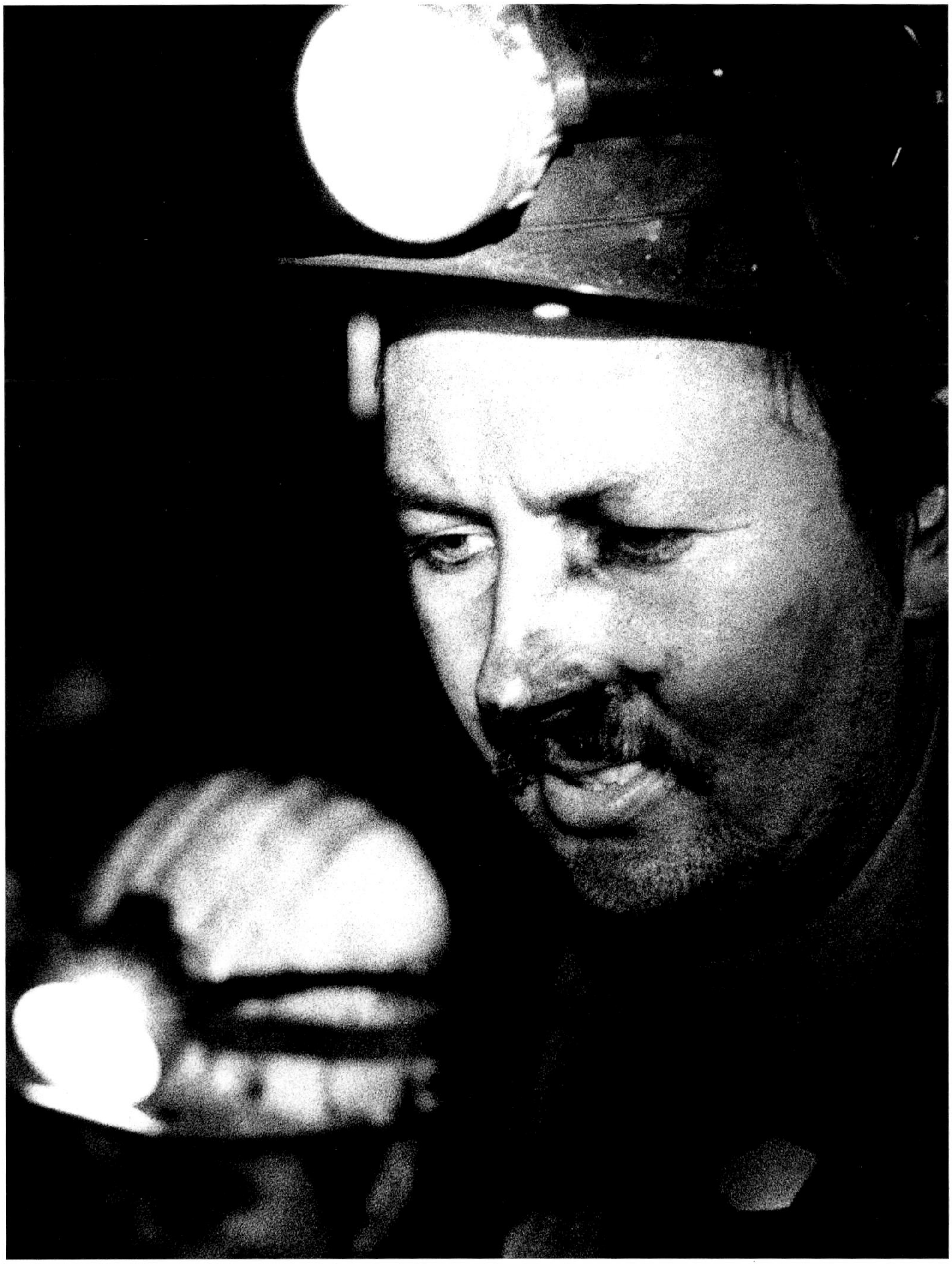

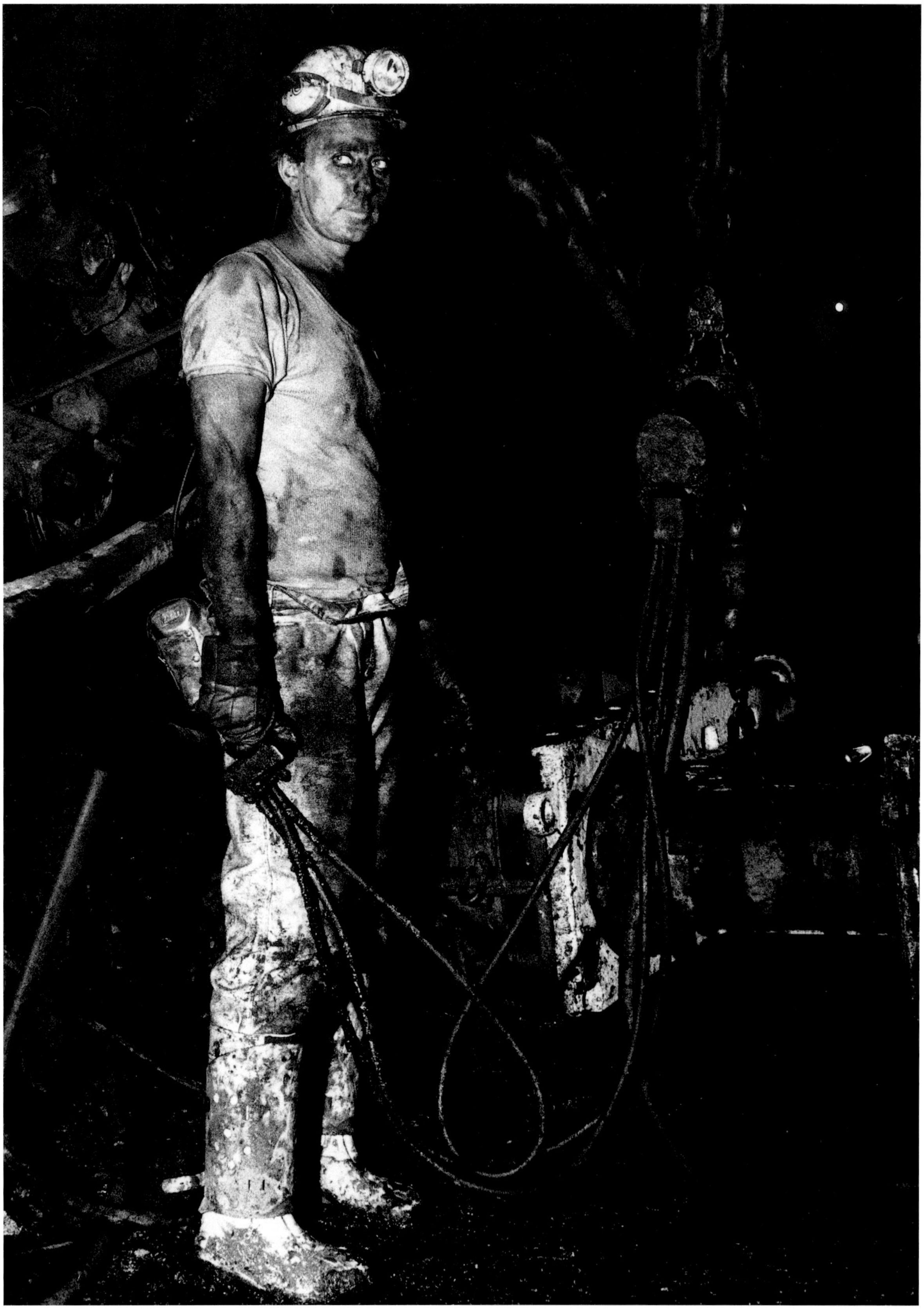

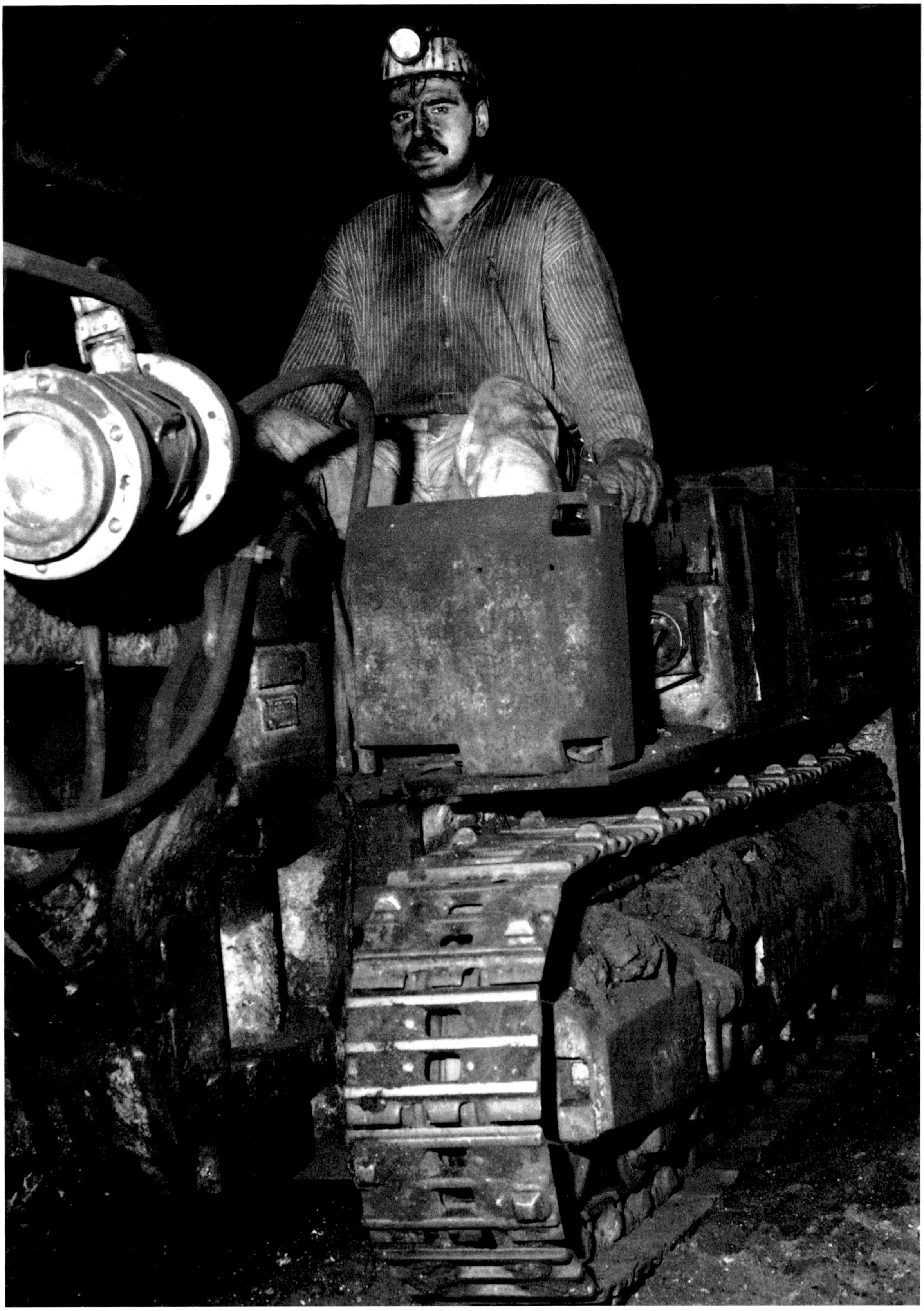

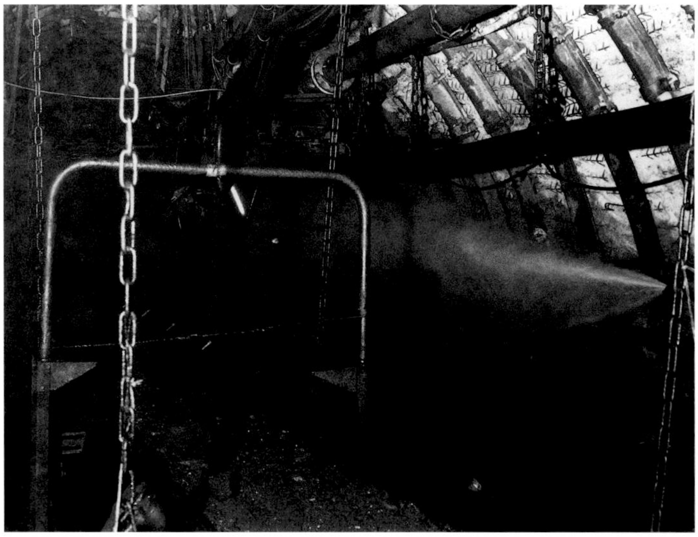

Coal is washed by a sprinkler system as it moves along the conveyor belt.

SHAFT LEVEL 203 MATHILDE, 28° C

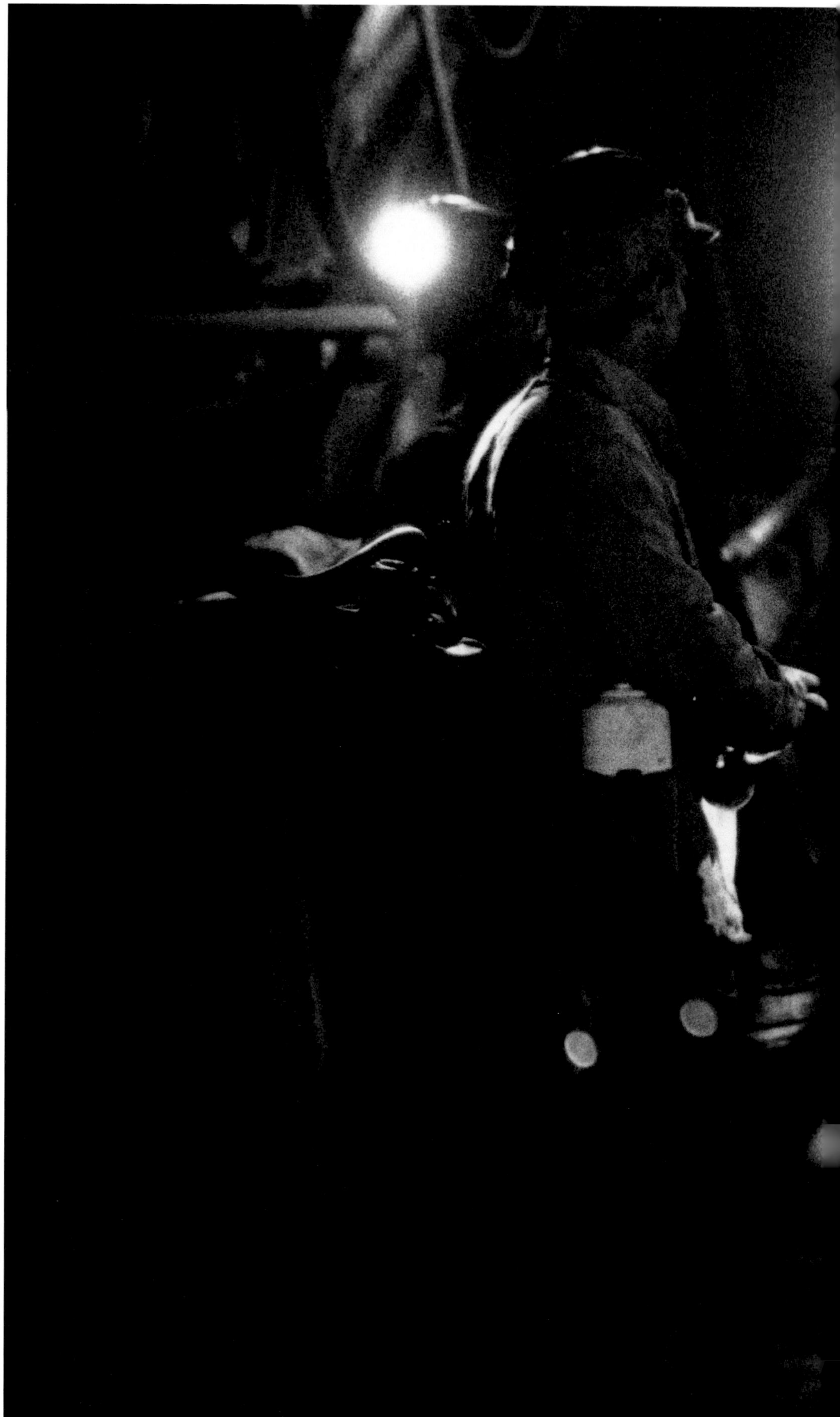

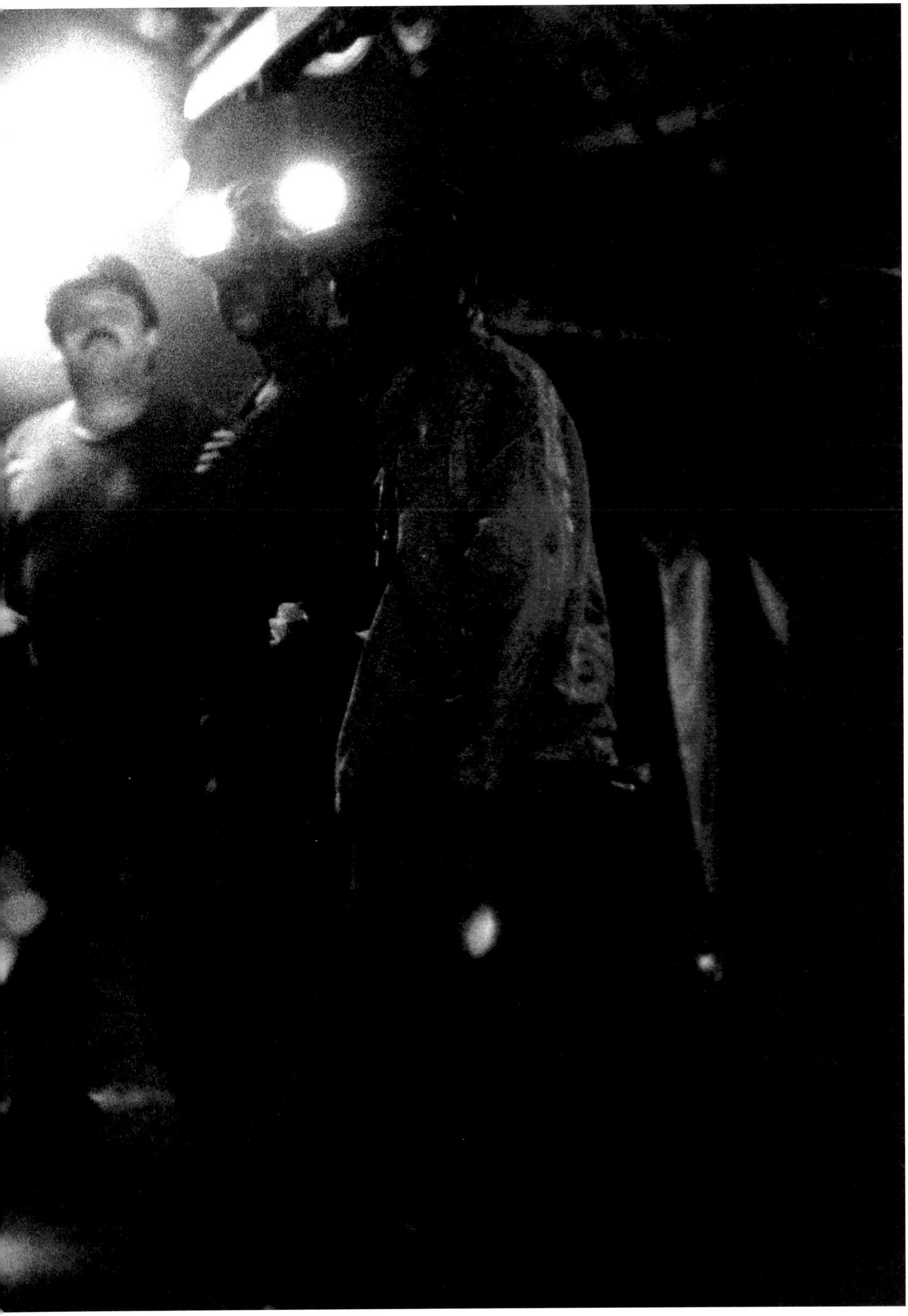

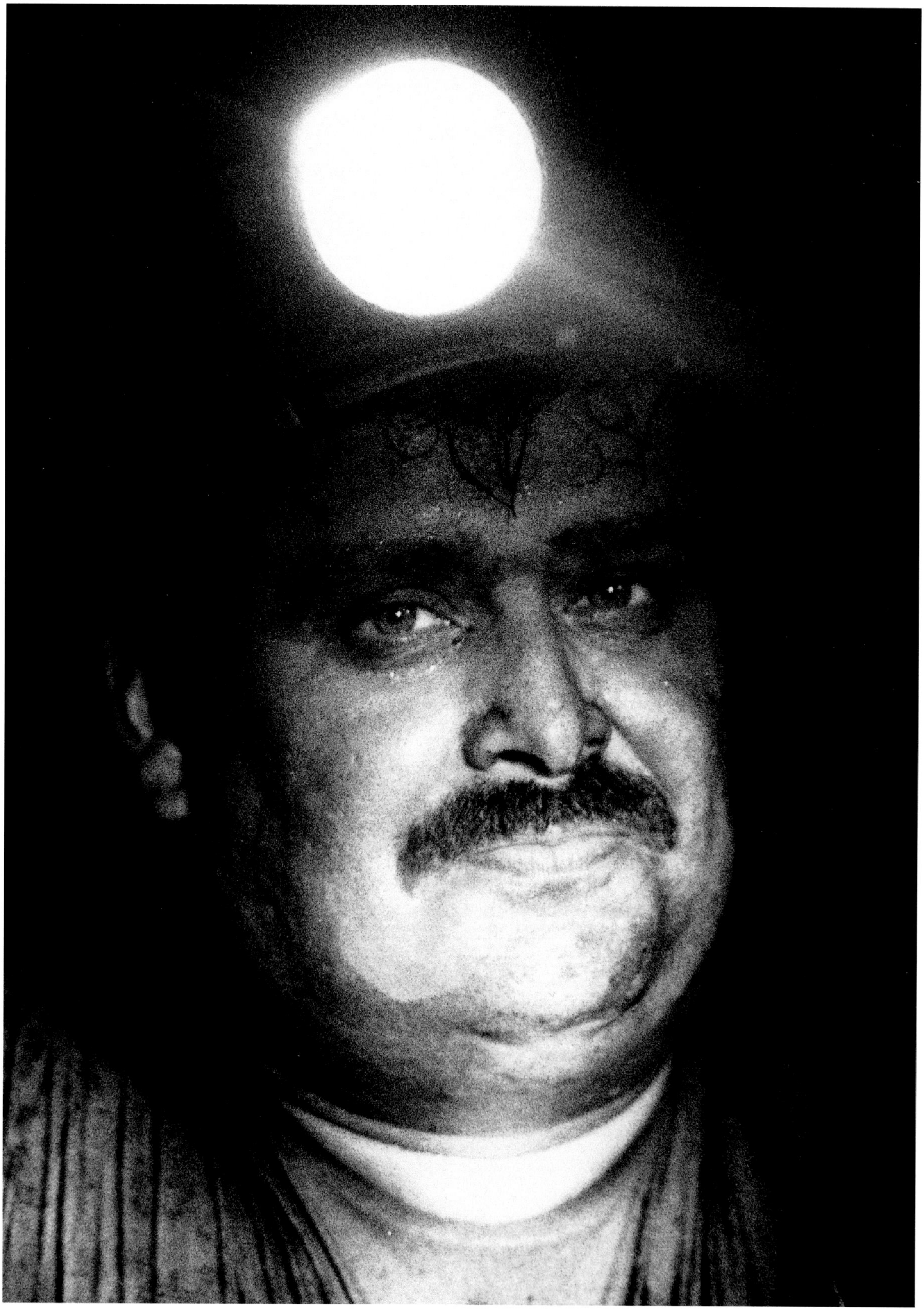

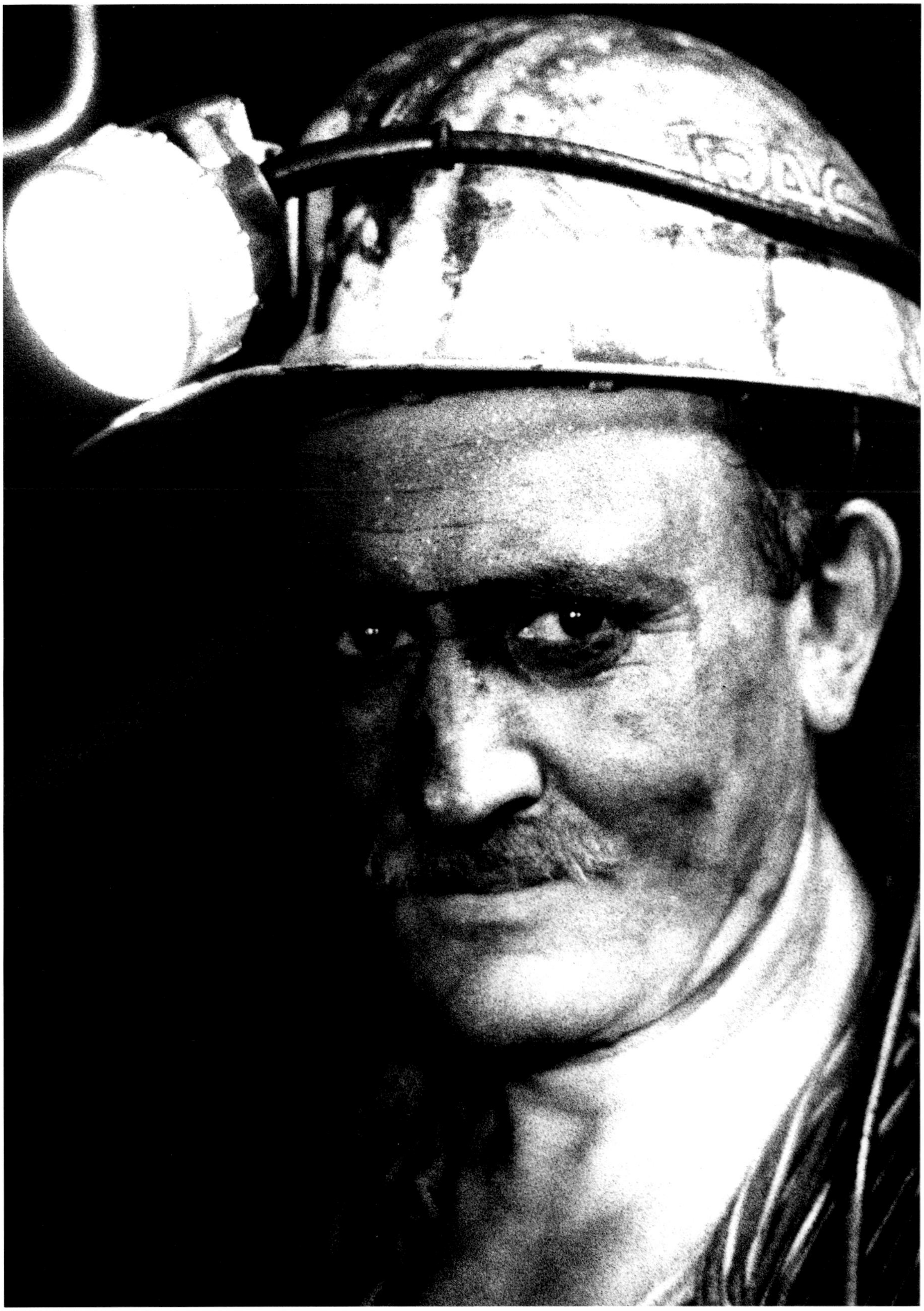

A **Drittel** or third – a team, ordinarily experienced in working together, which mans a specific shift.

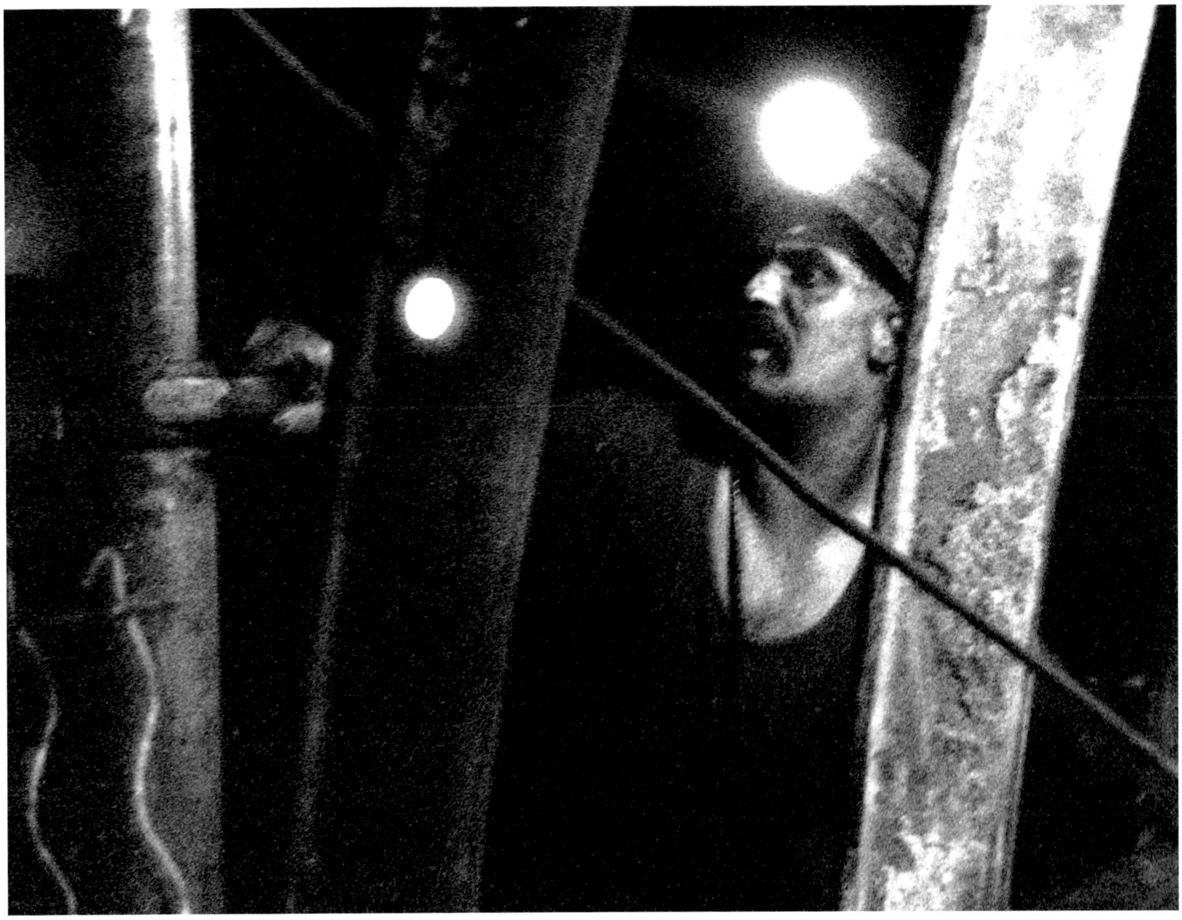

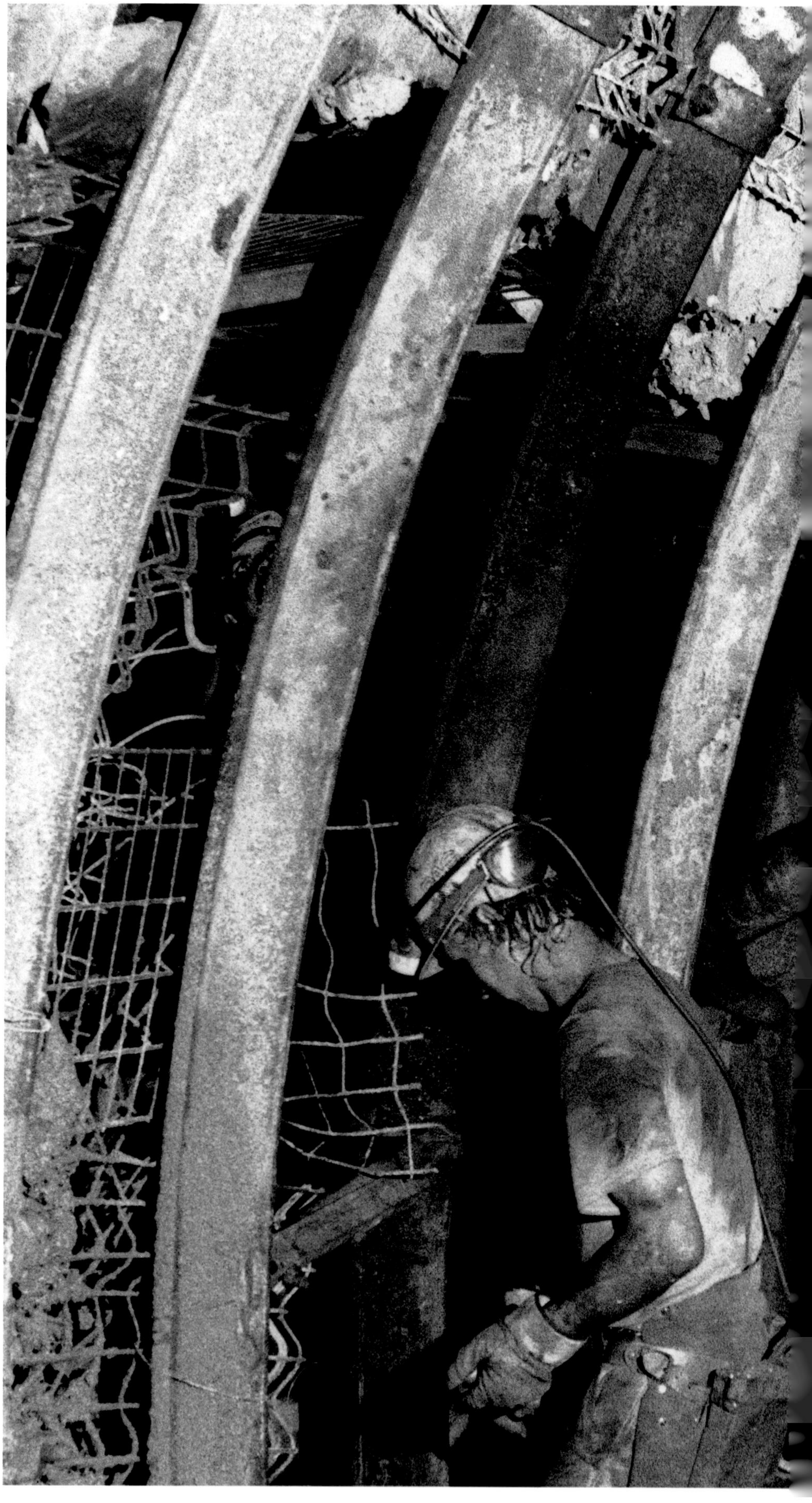

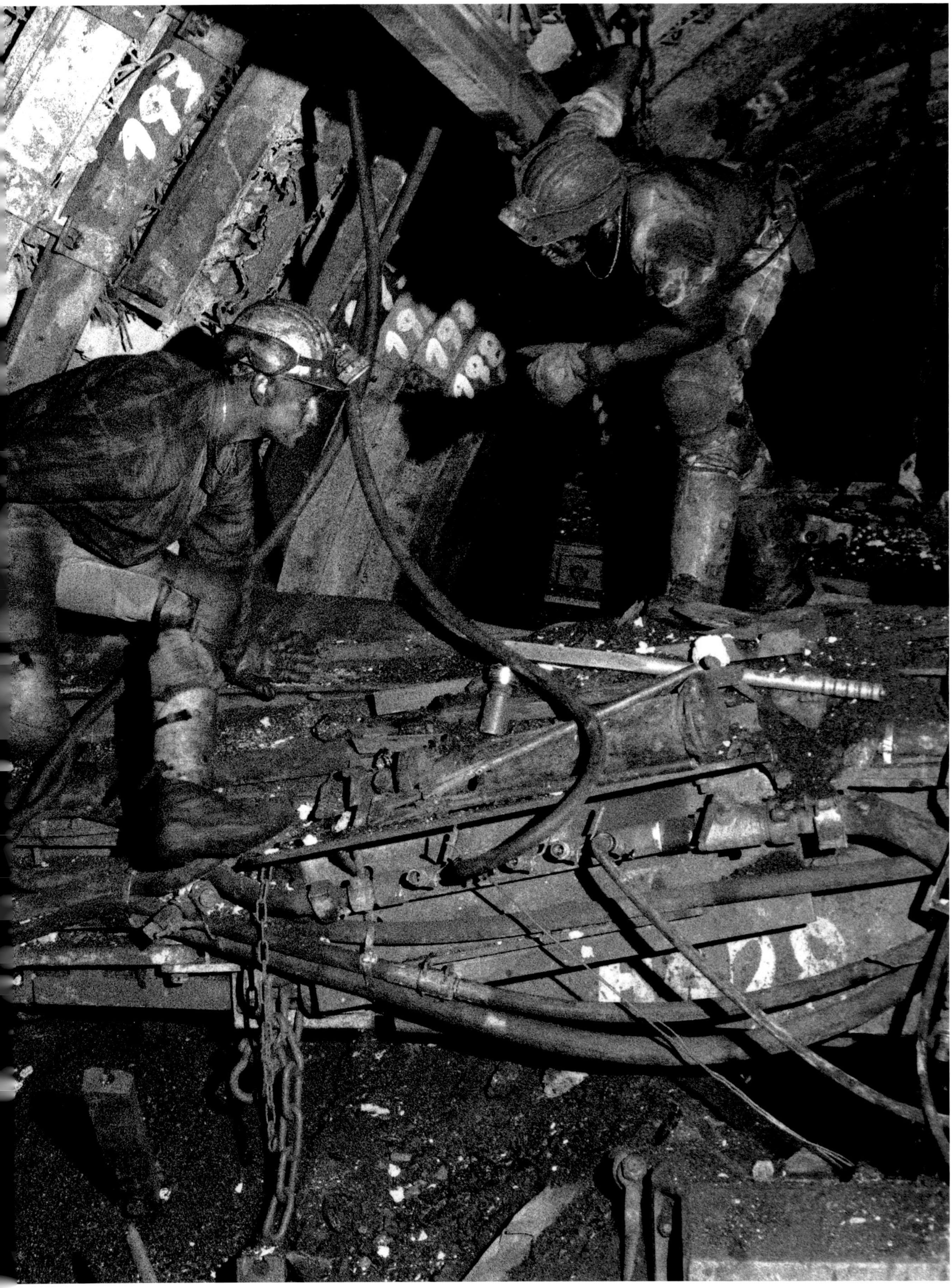

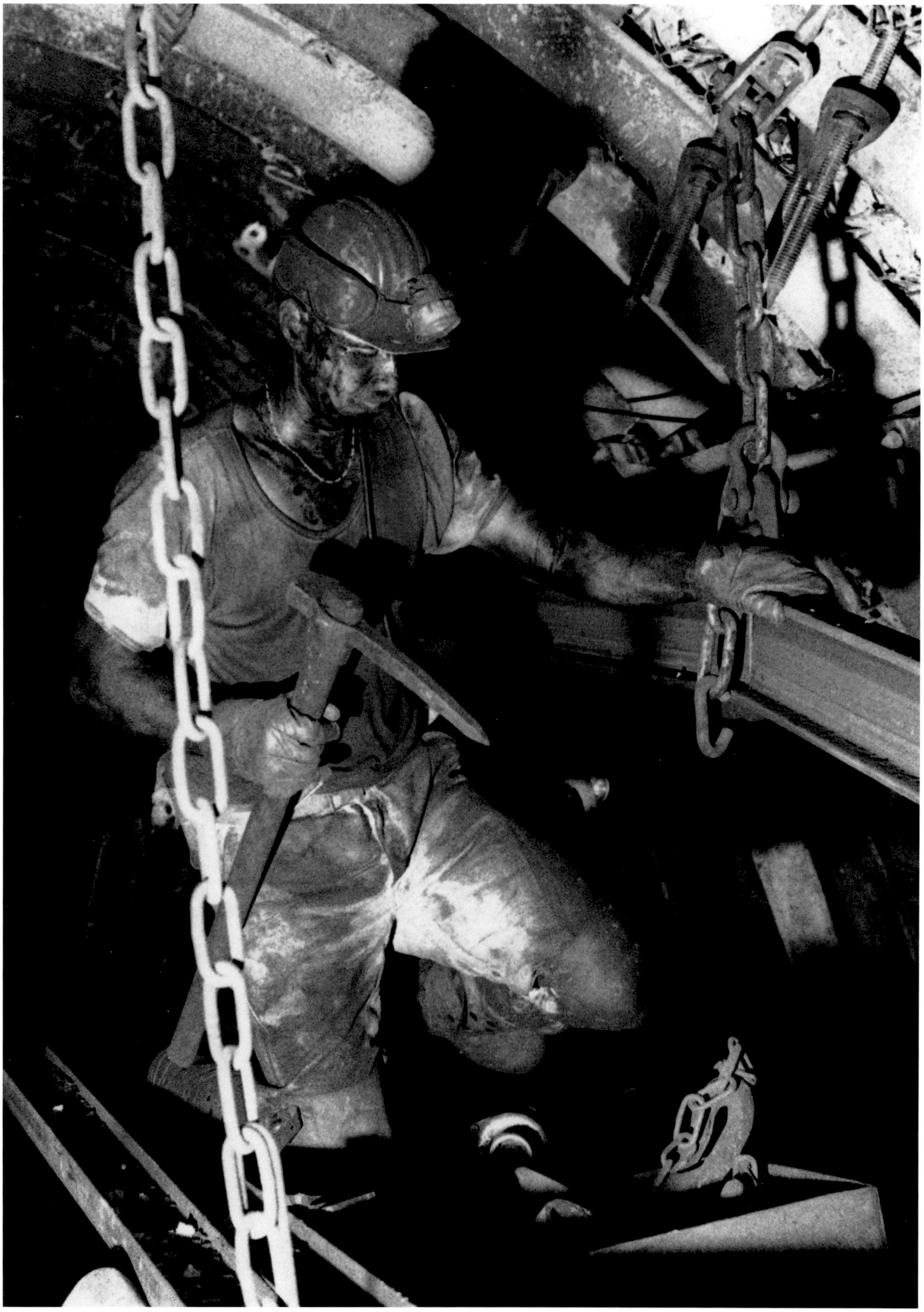

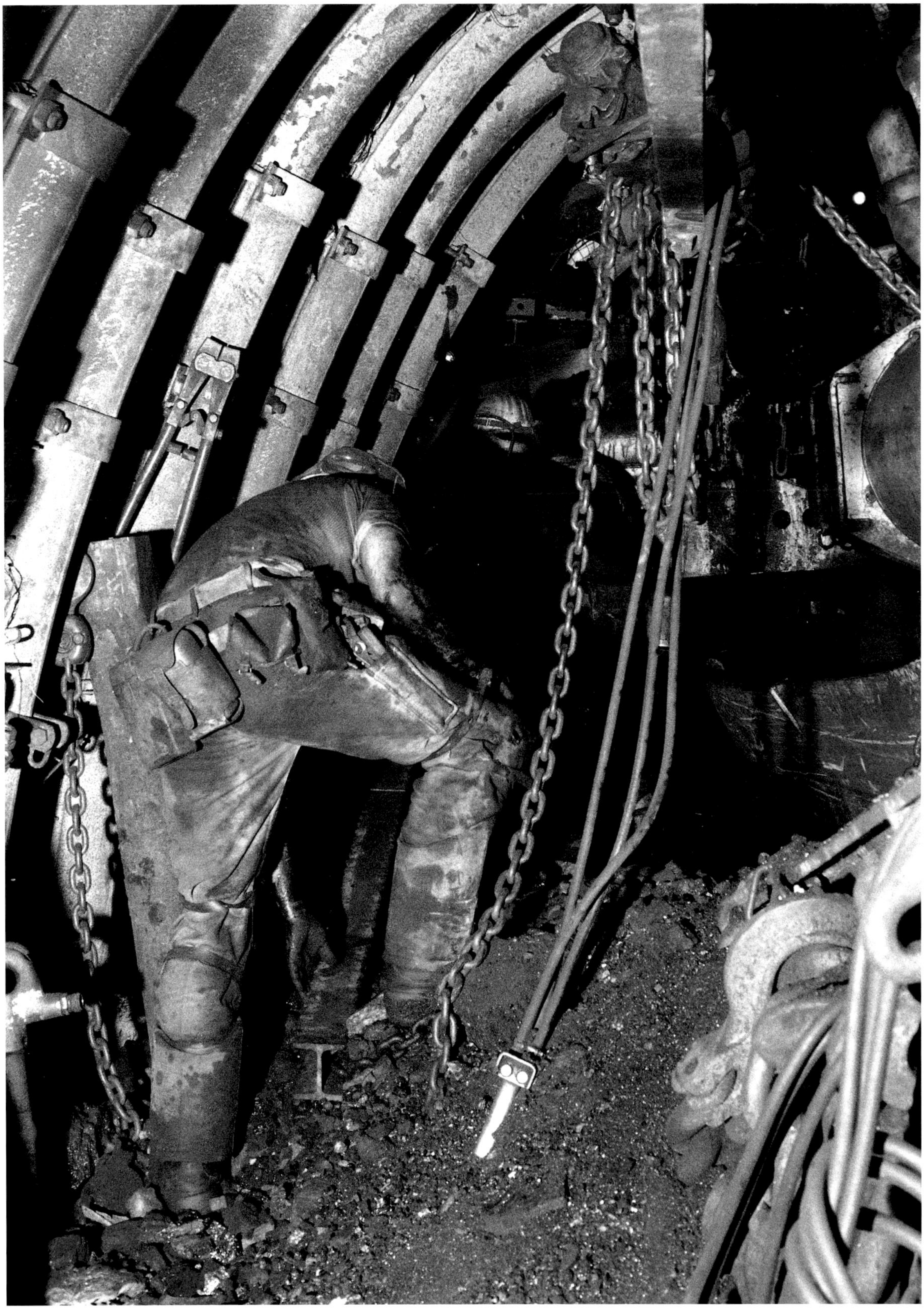

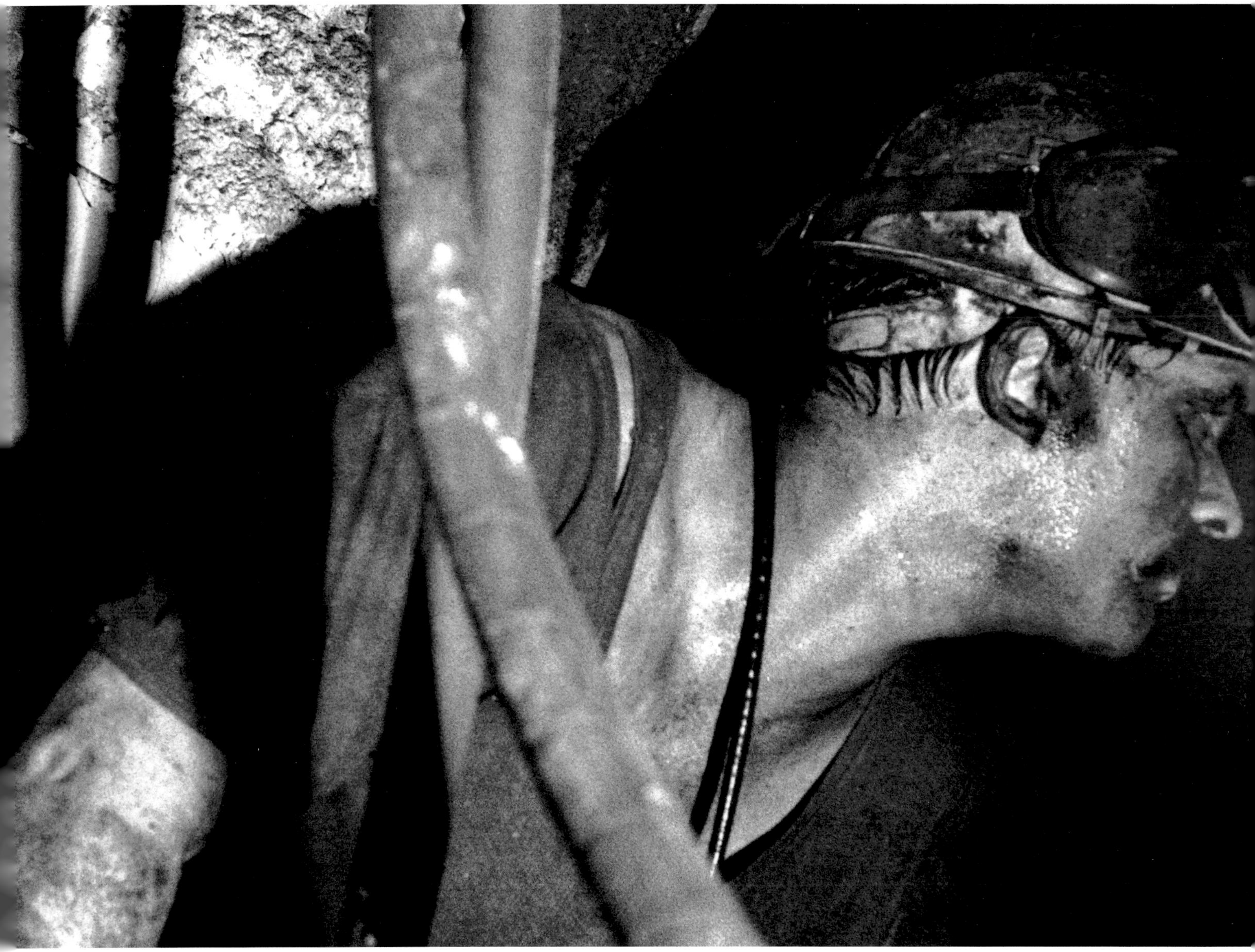

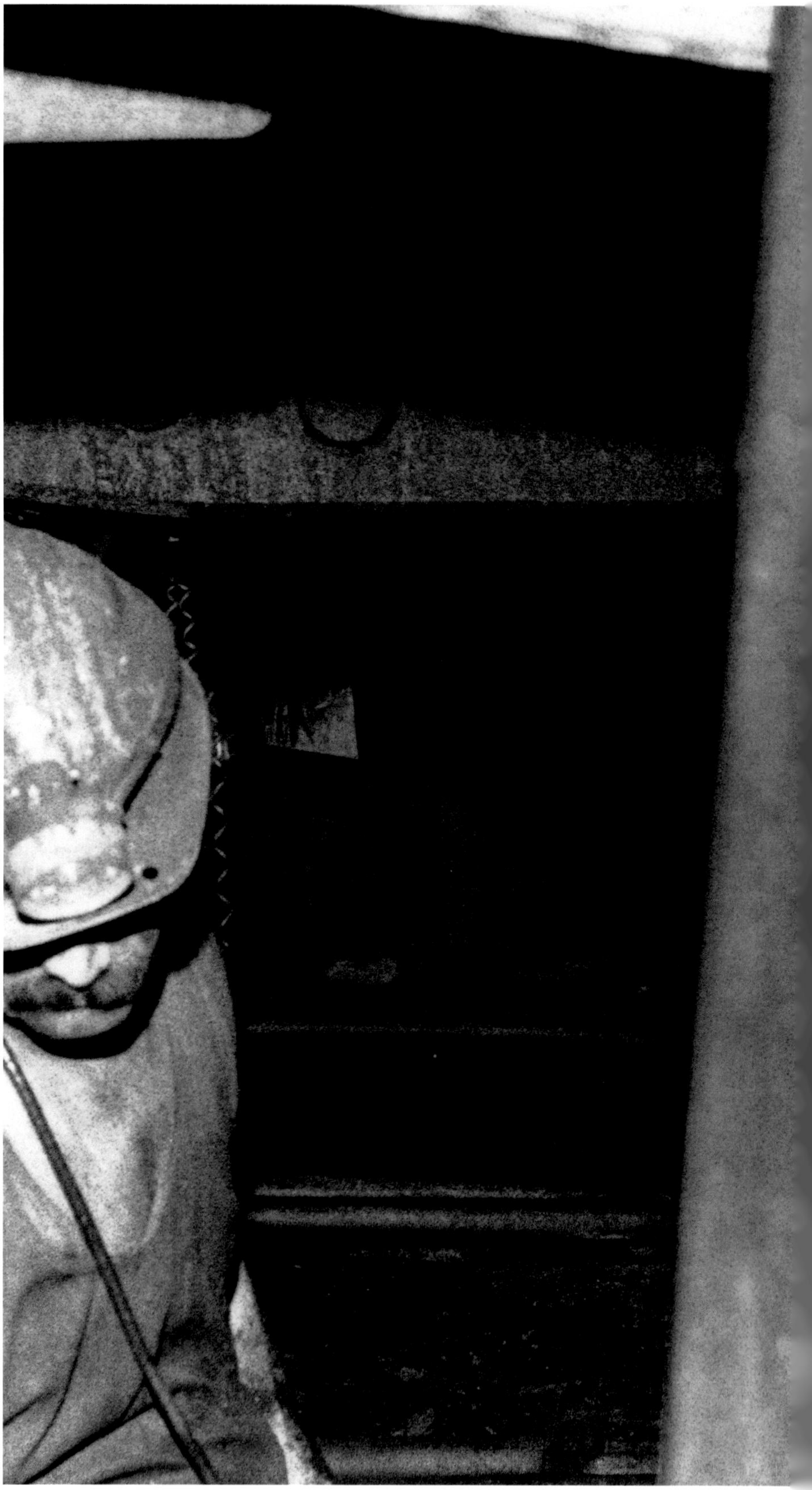

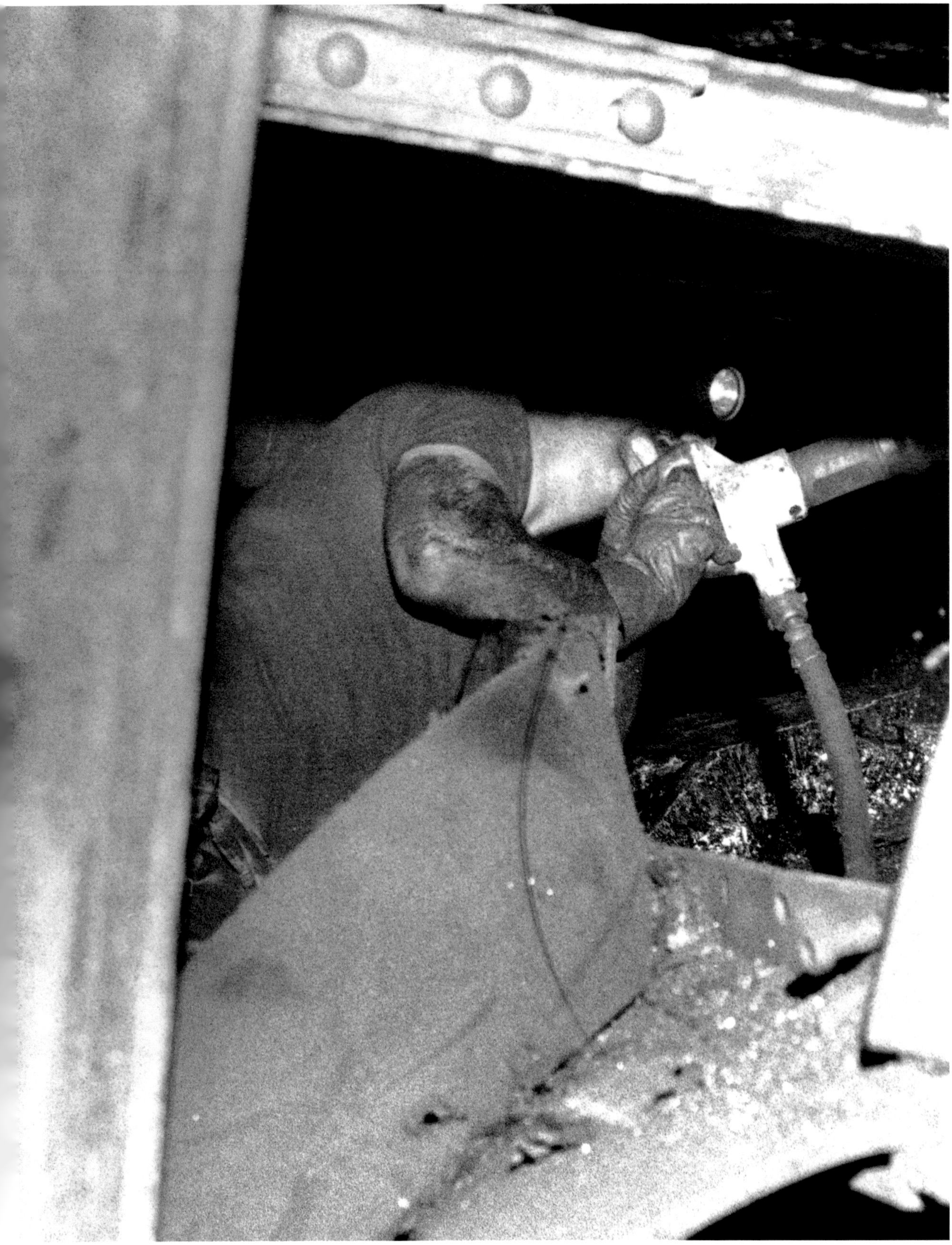

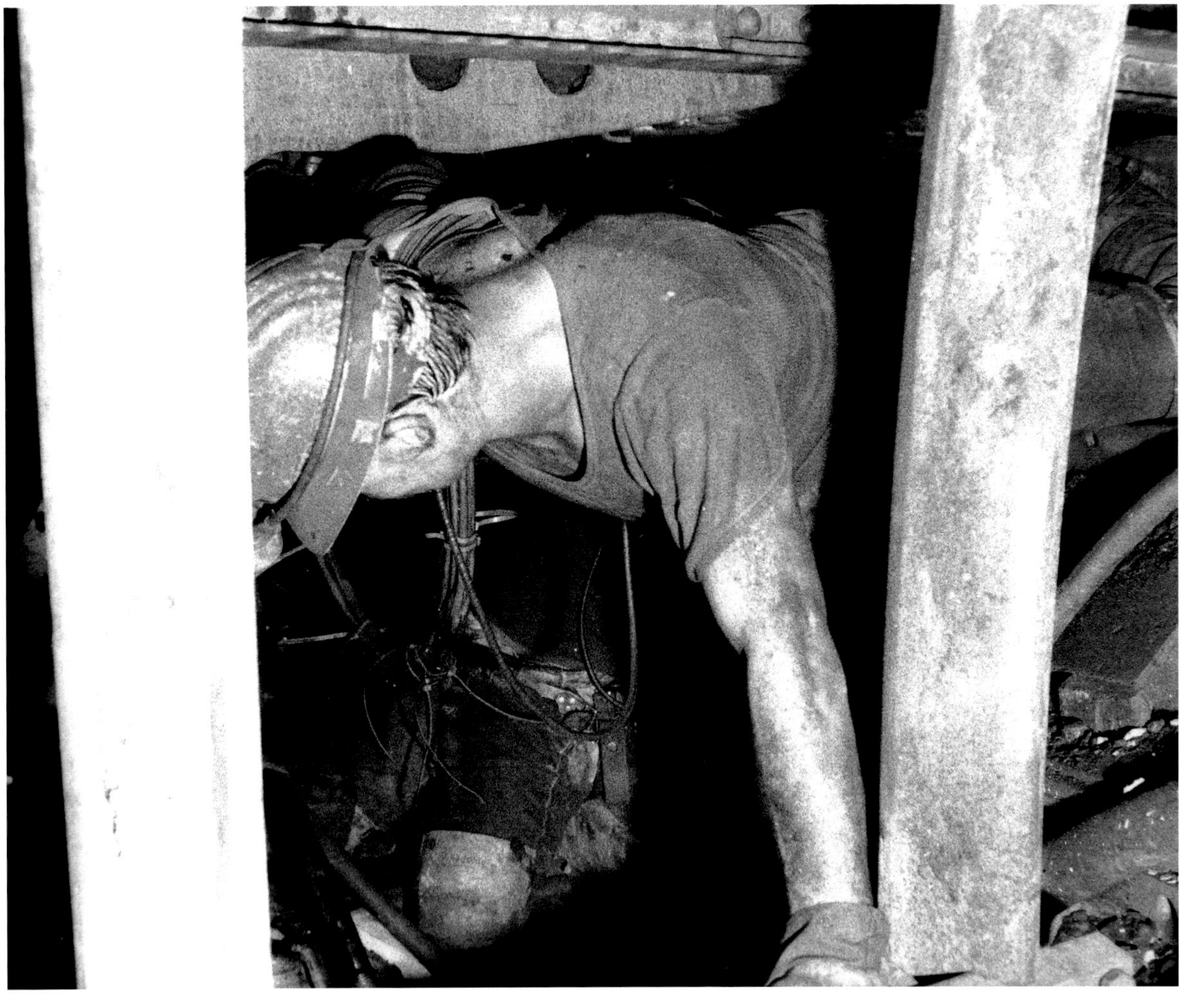

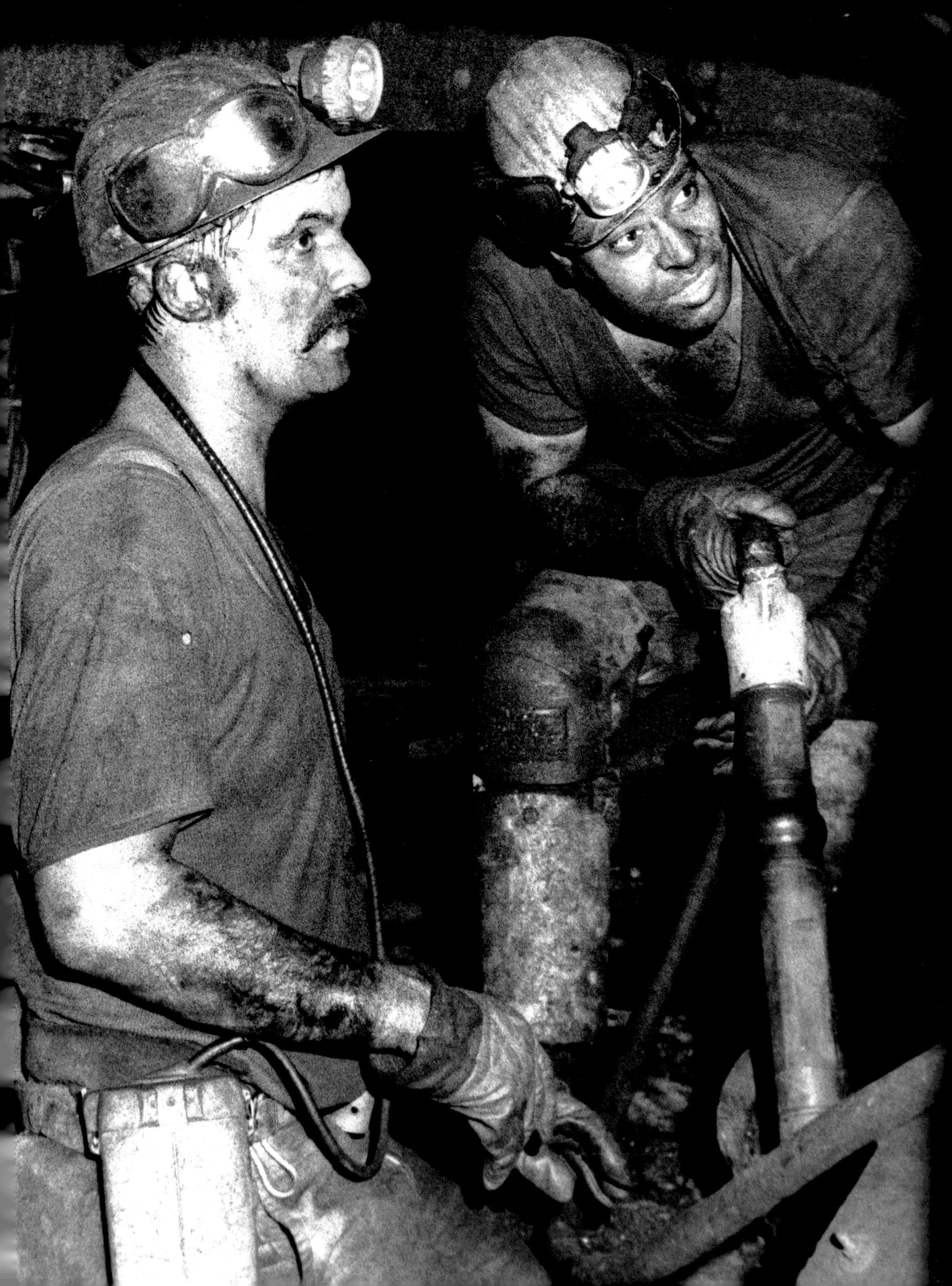

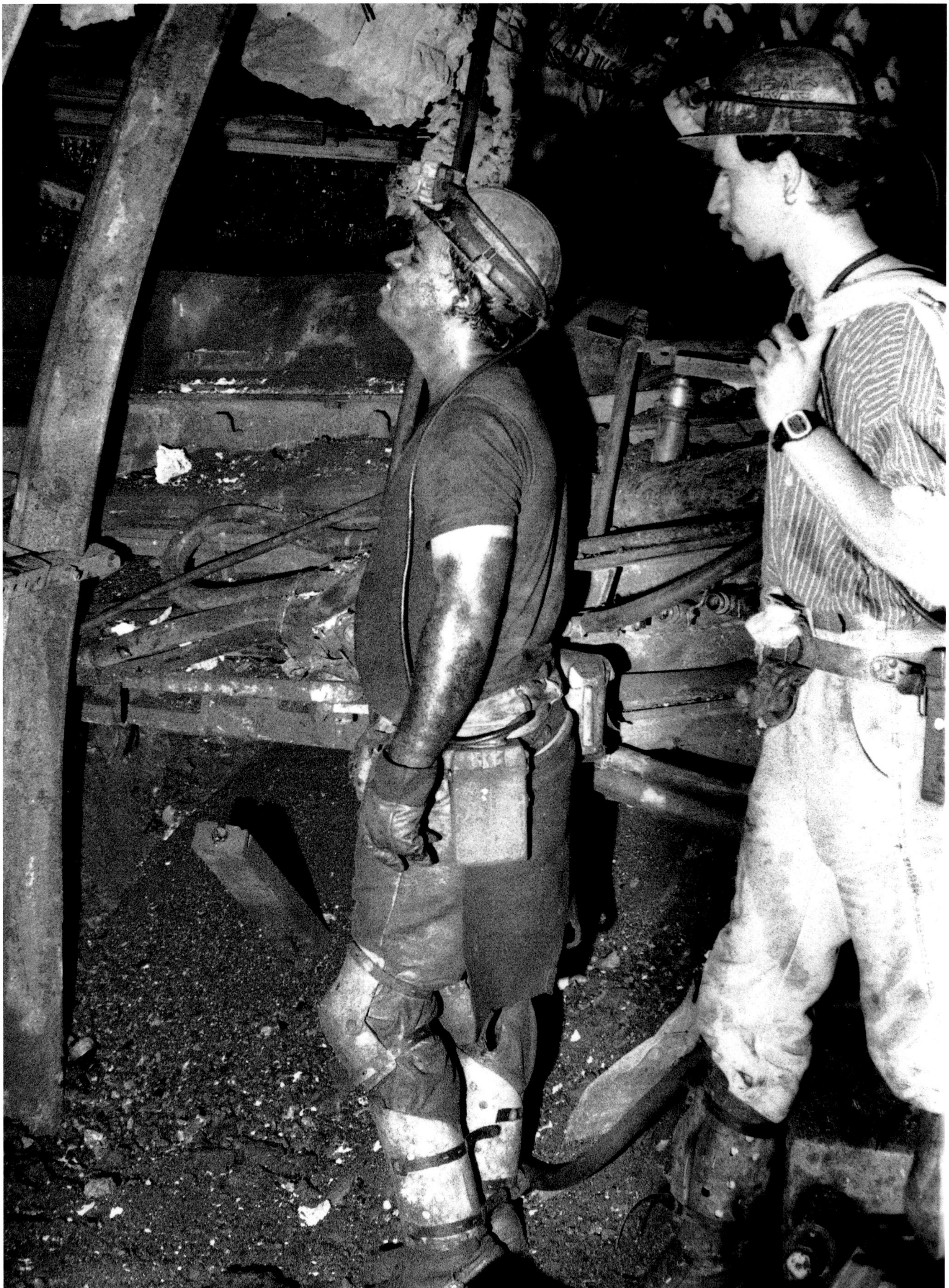

Performance-based pay – pay dependent on work production. Mining law permits the "Gedinge," a typical form of compensation for miners who work underground.

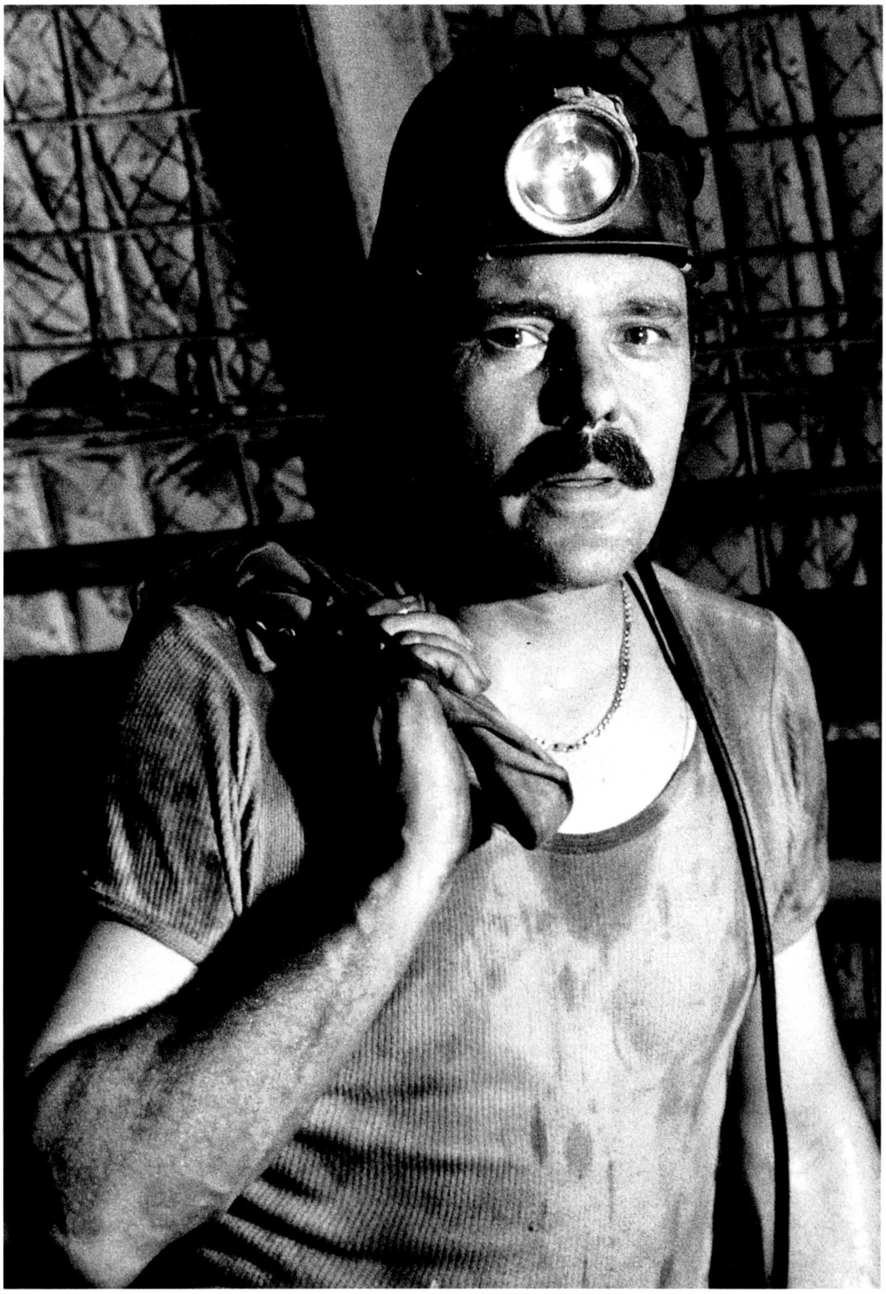

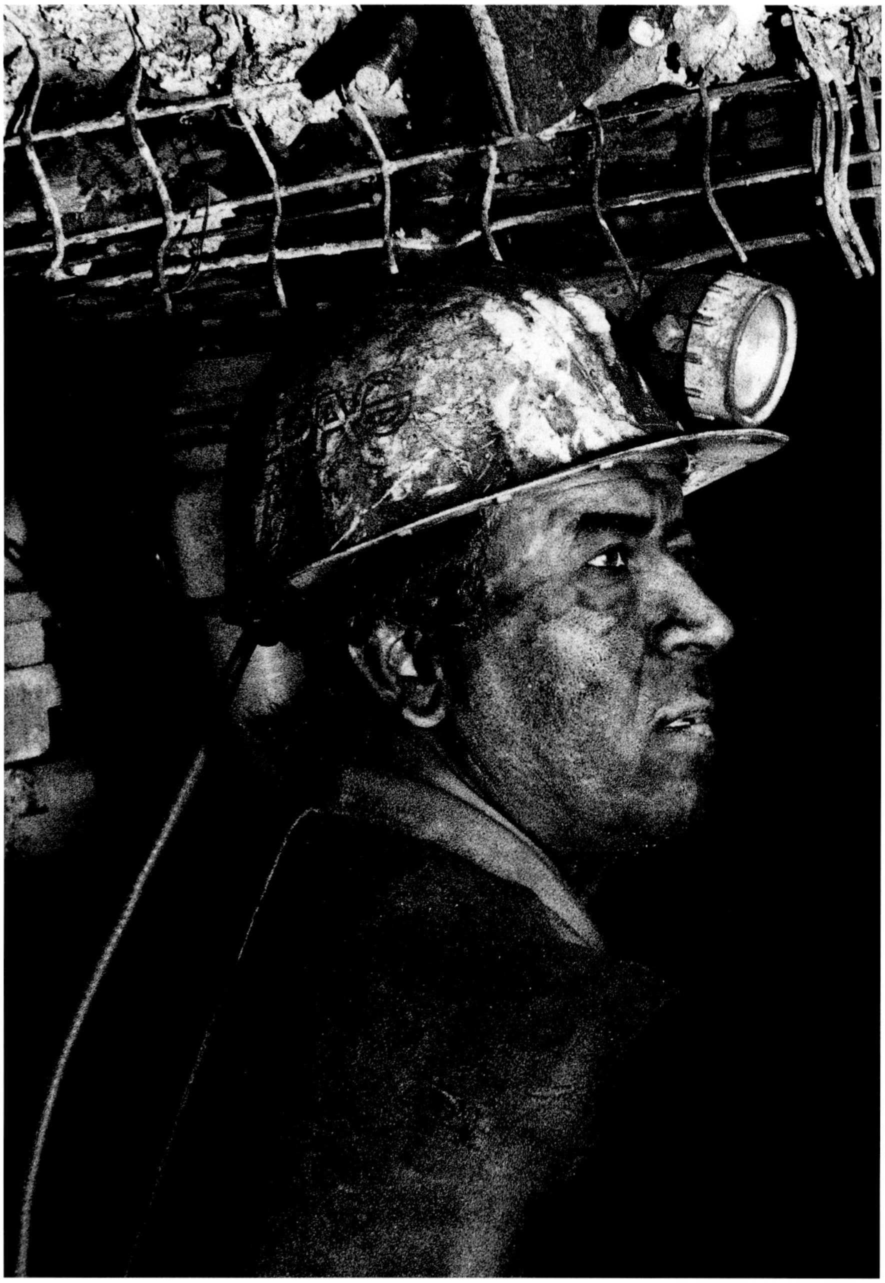

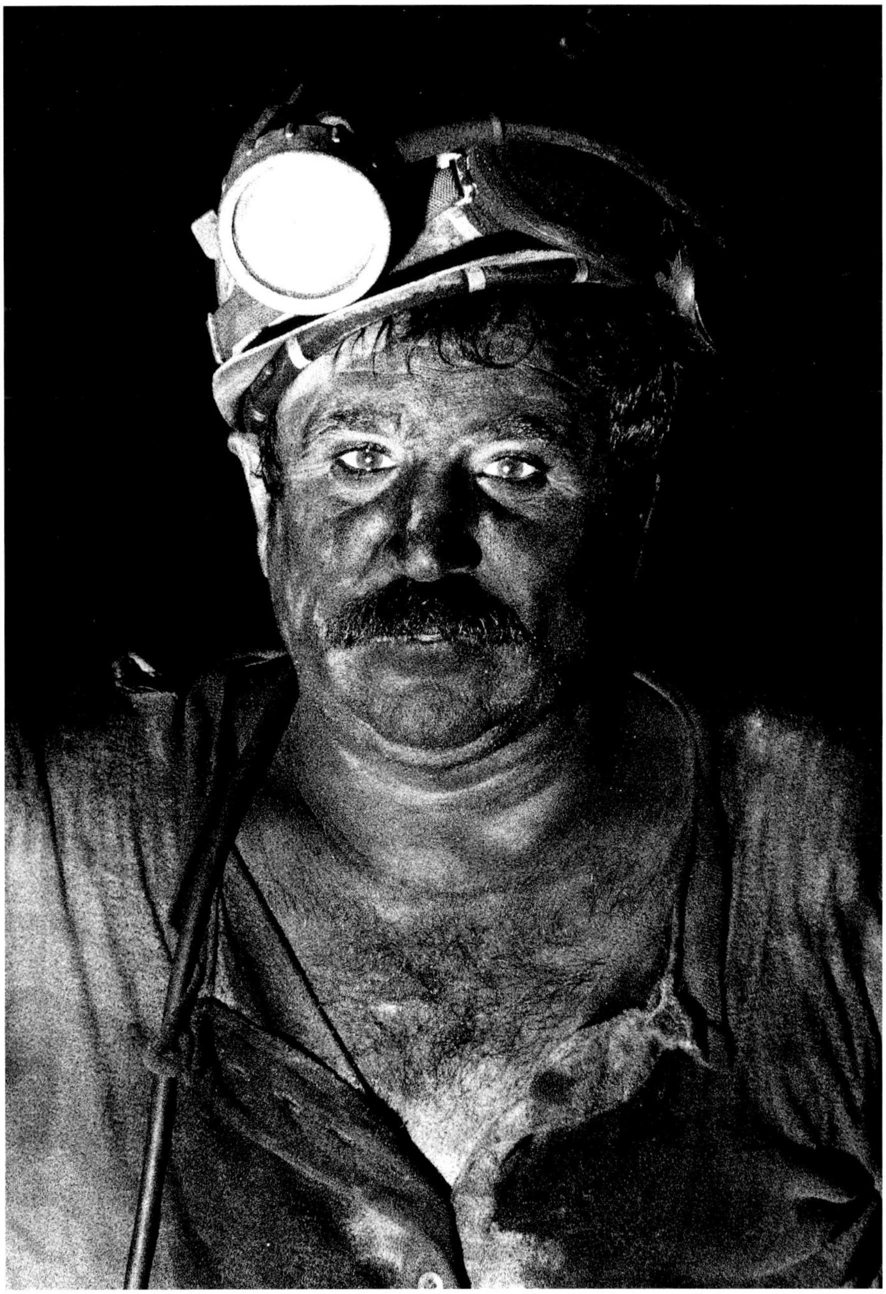

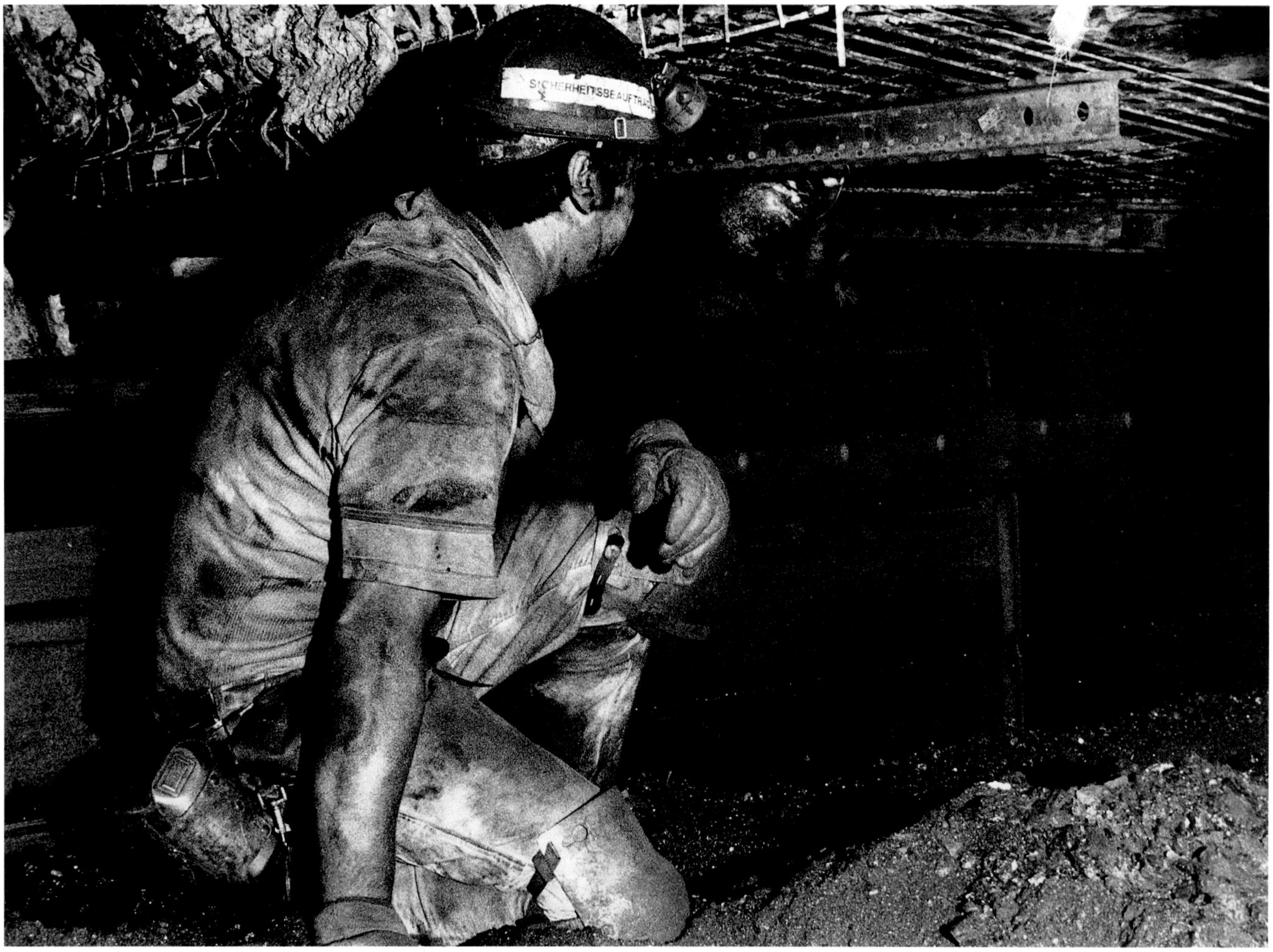
SICHERHEITSBEAUFTRA

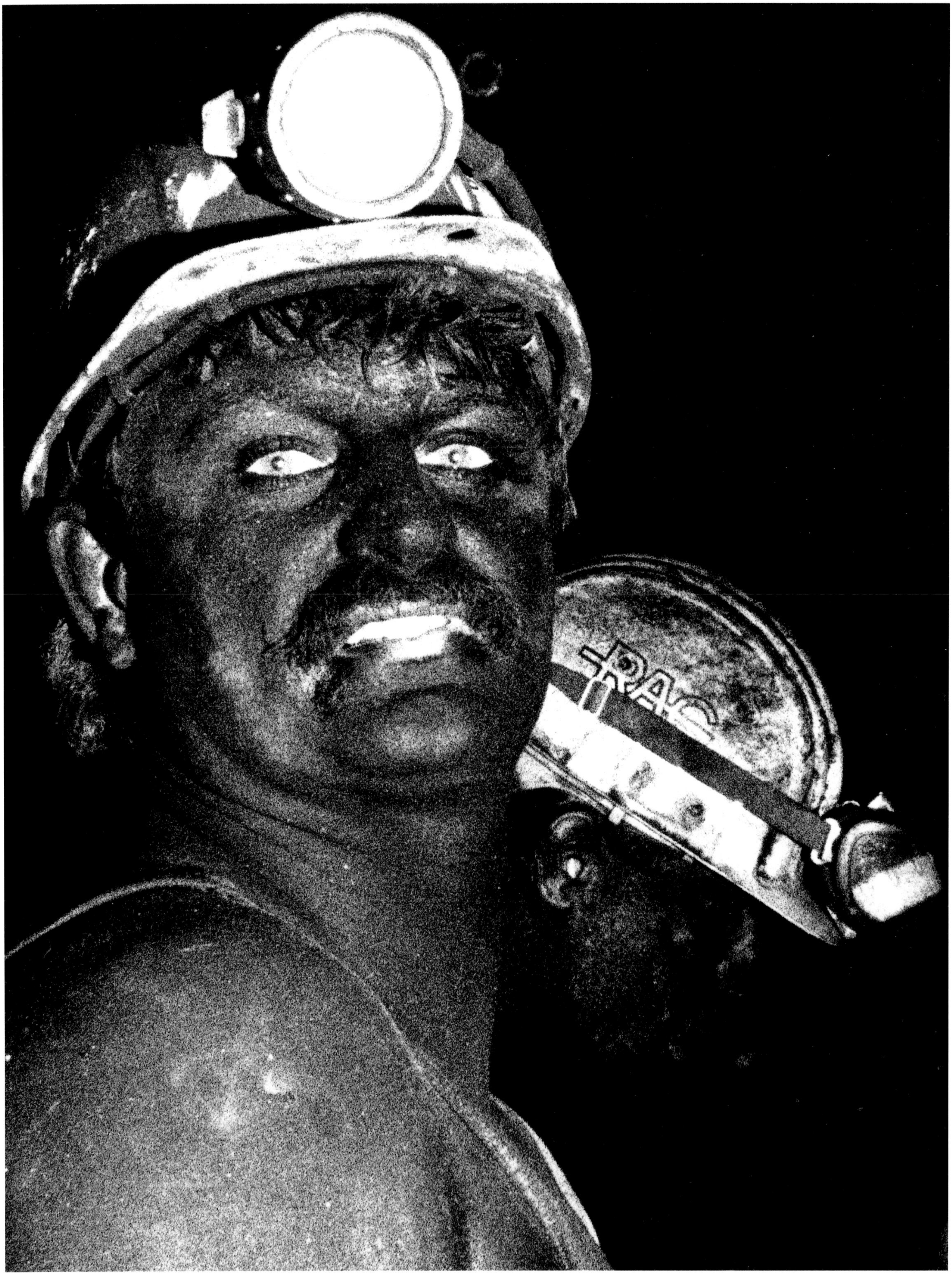

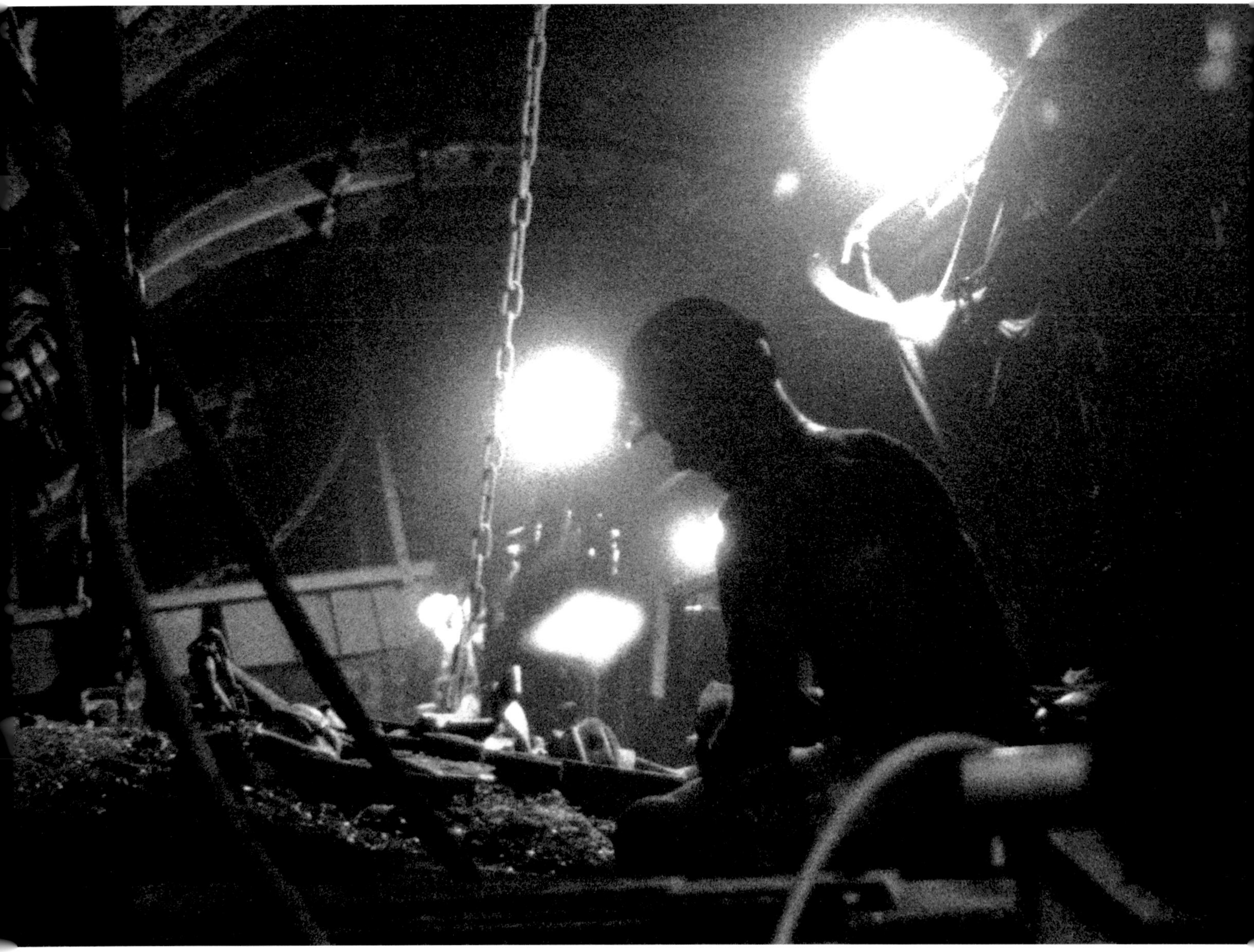

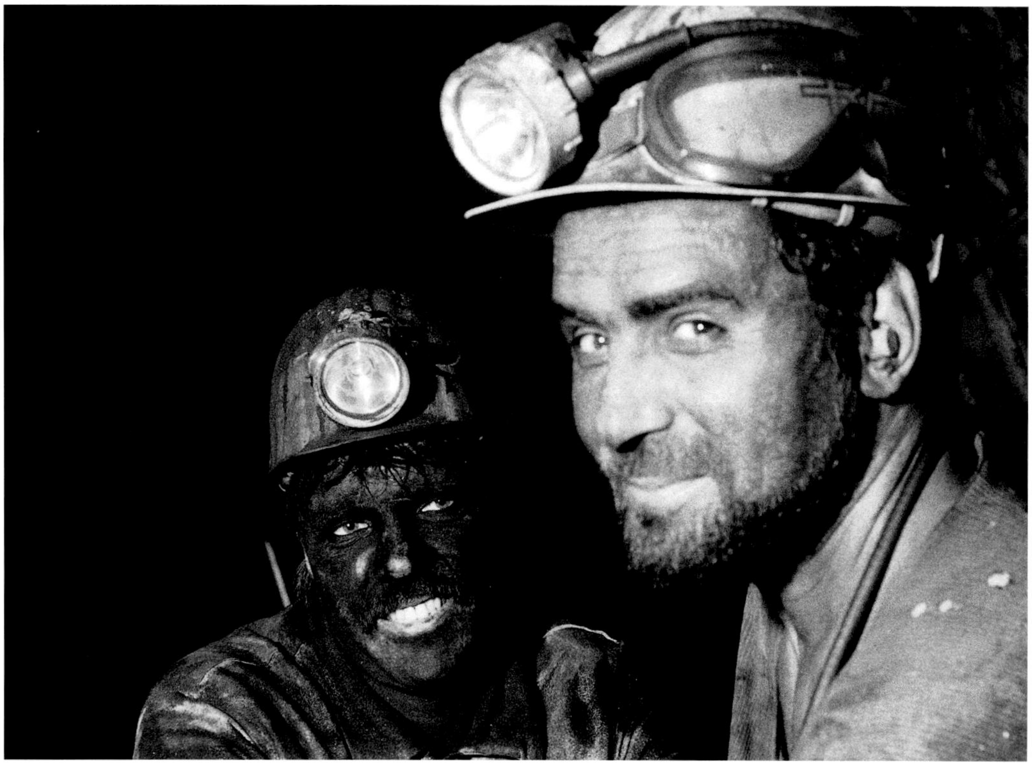

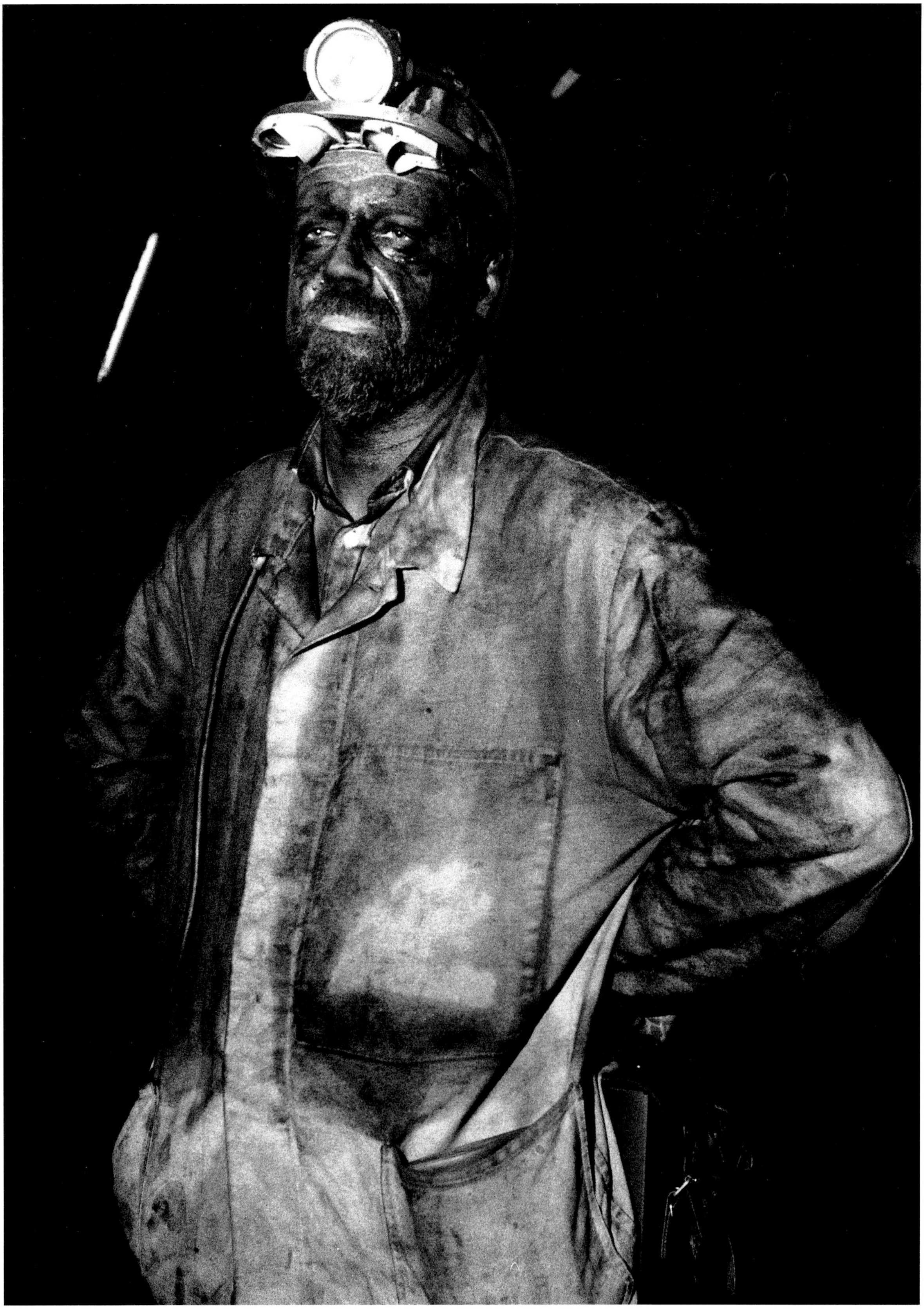

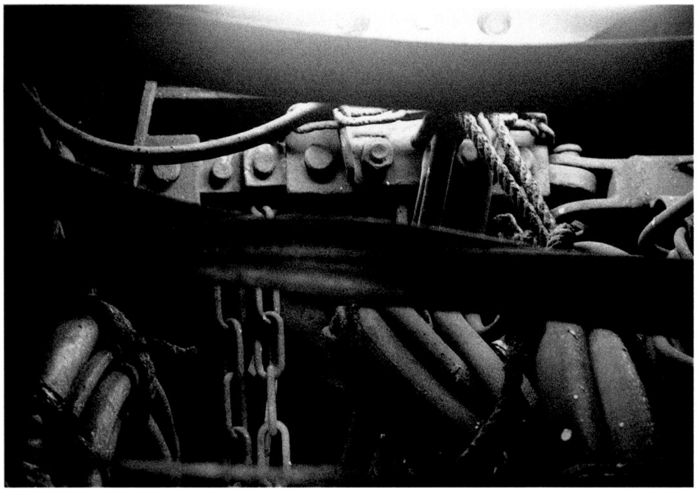

Cable trunk in the coal output line

SHAFT LEVEL 478 KATHARINA, 31° C

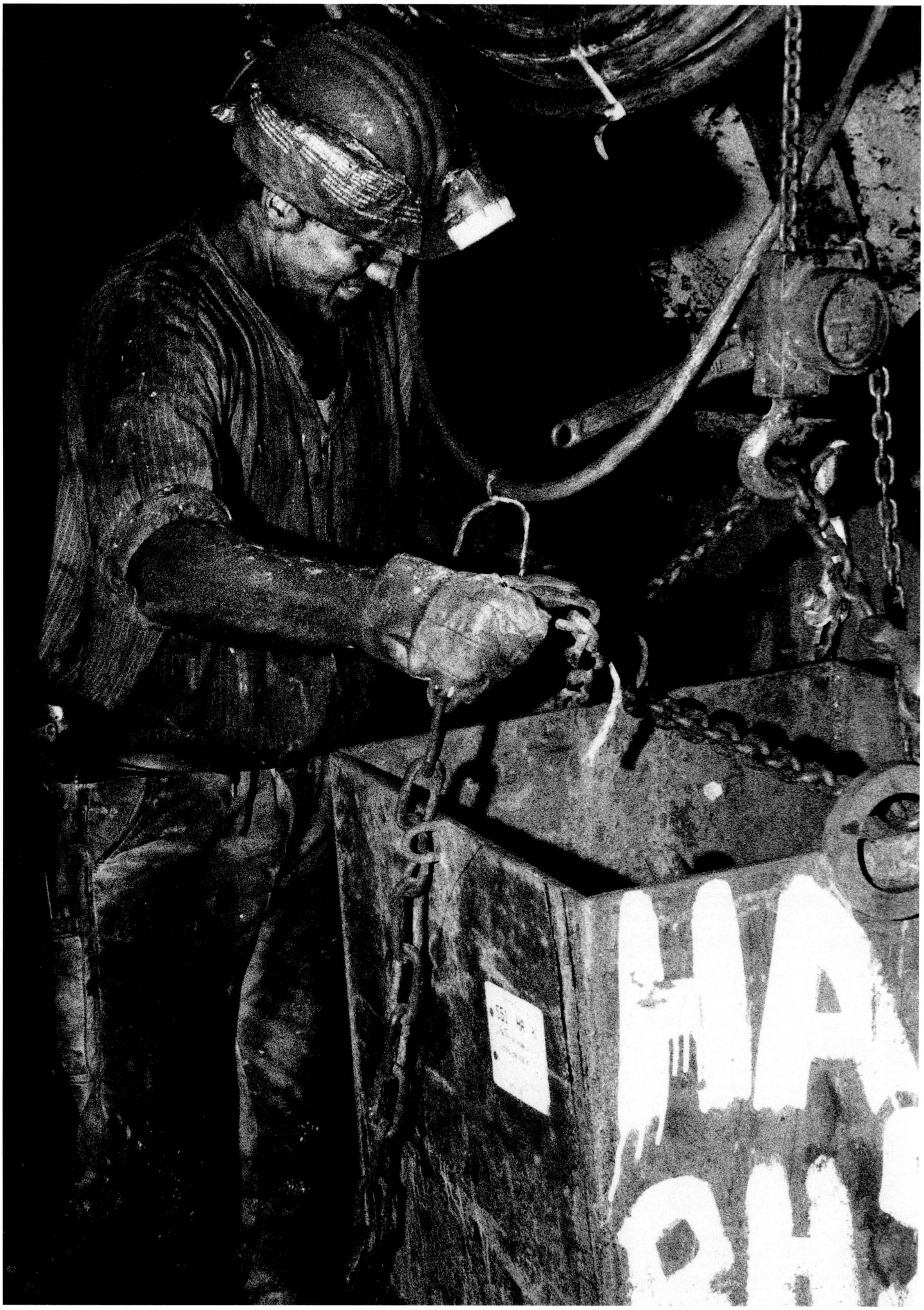

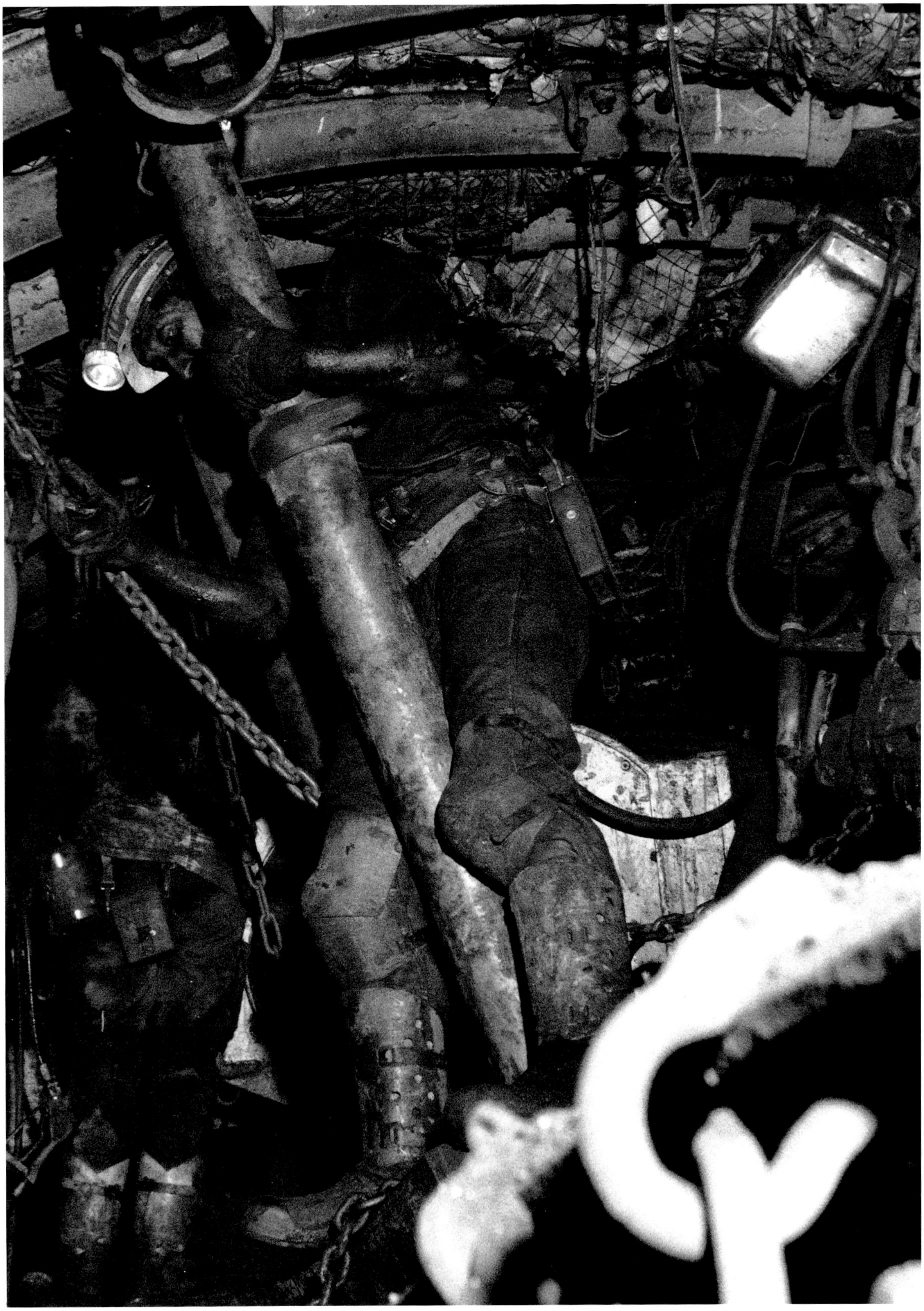

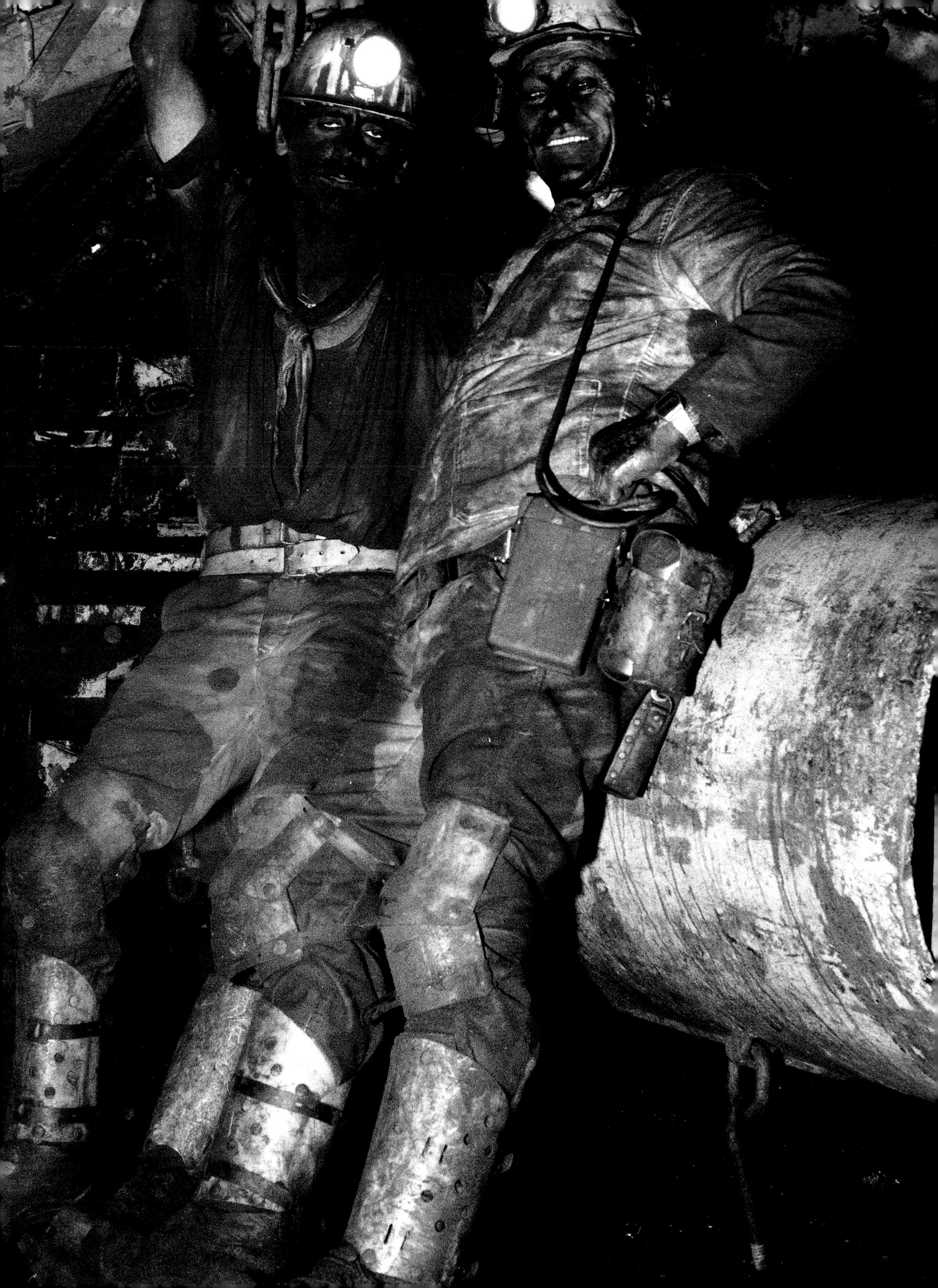

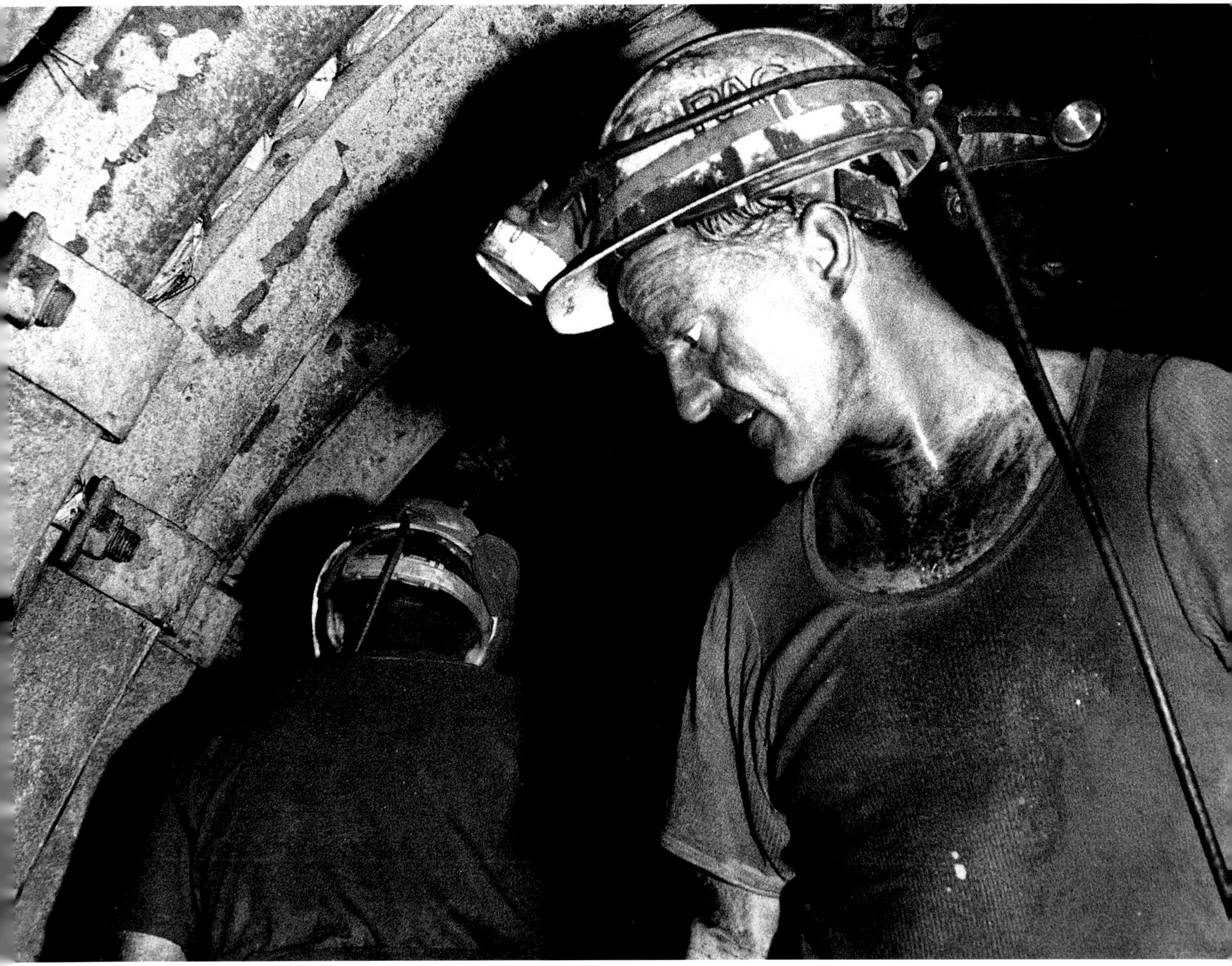

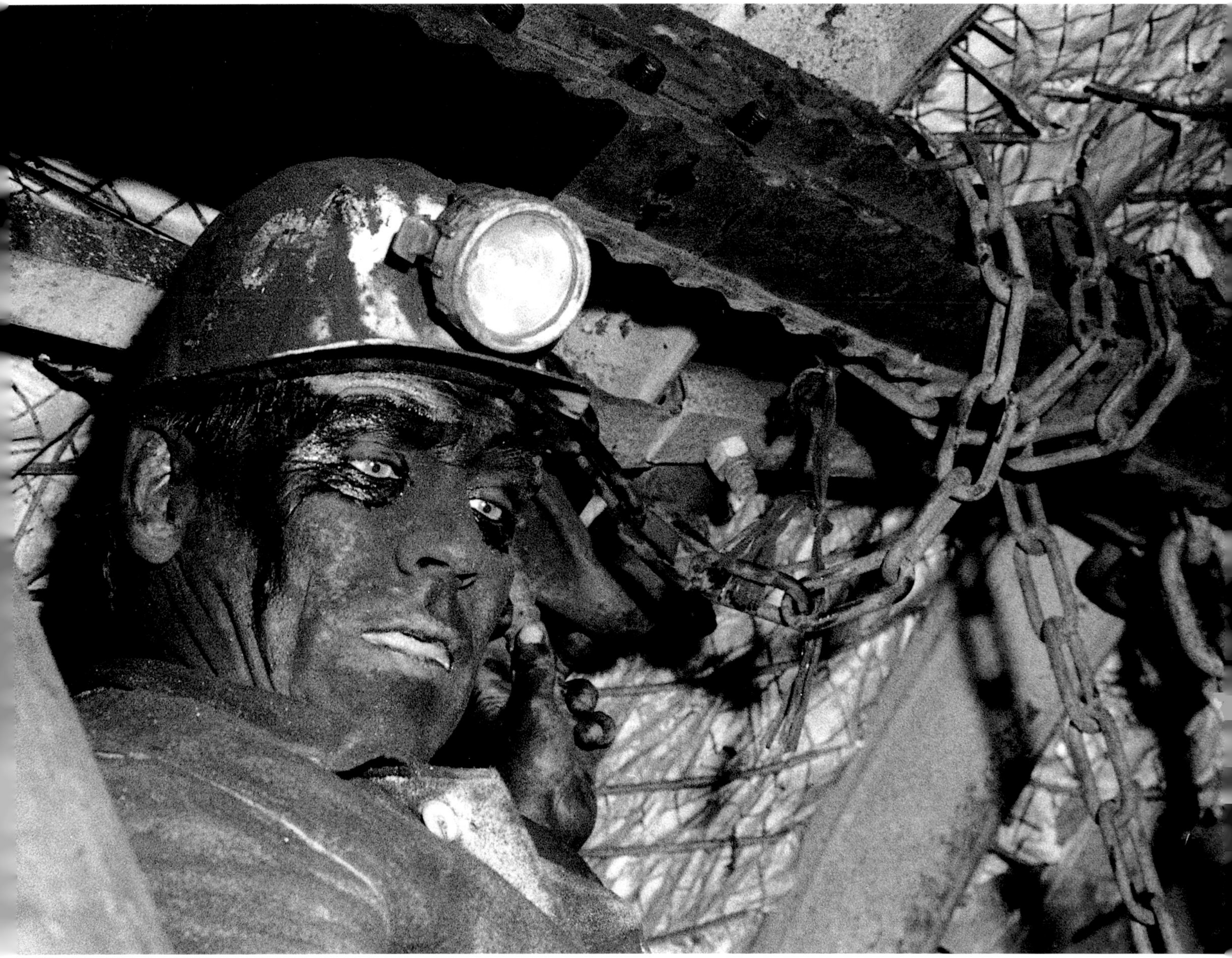

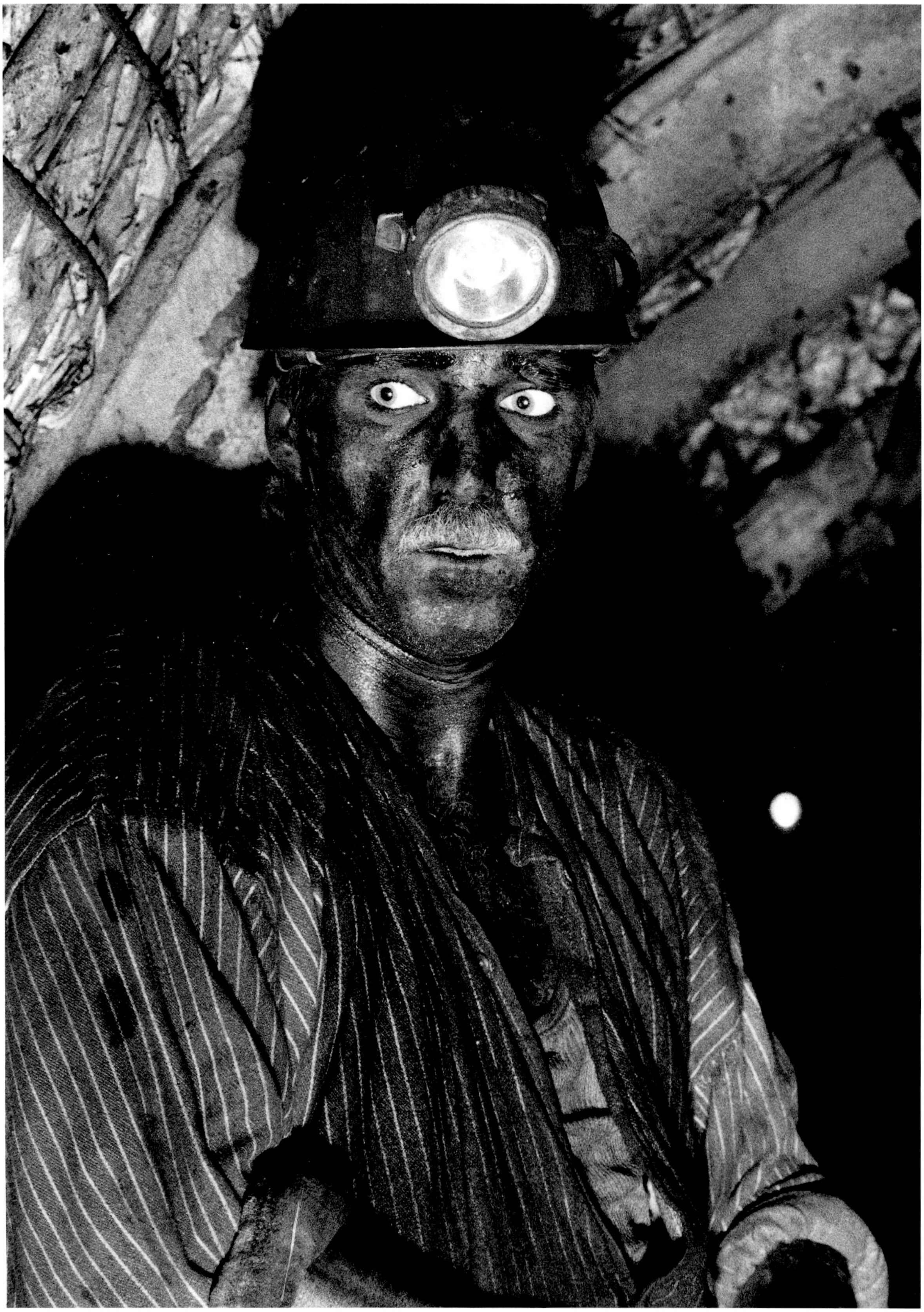

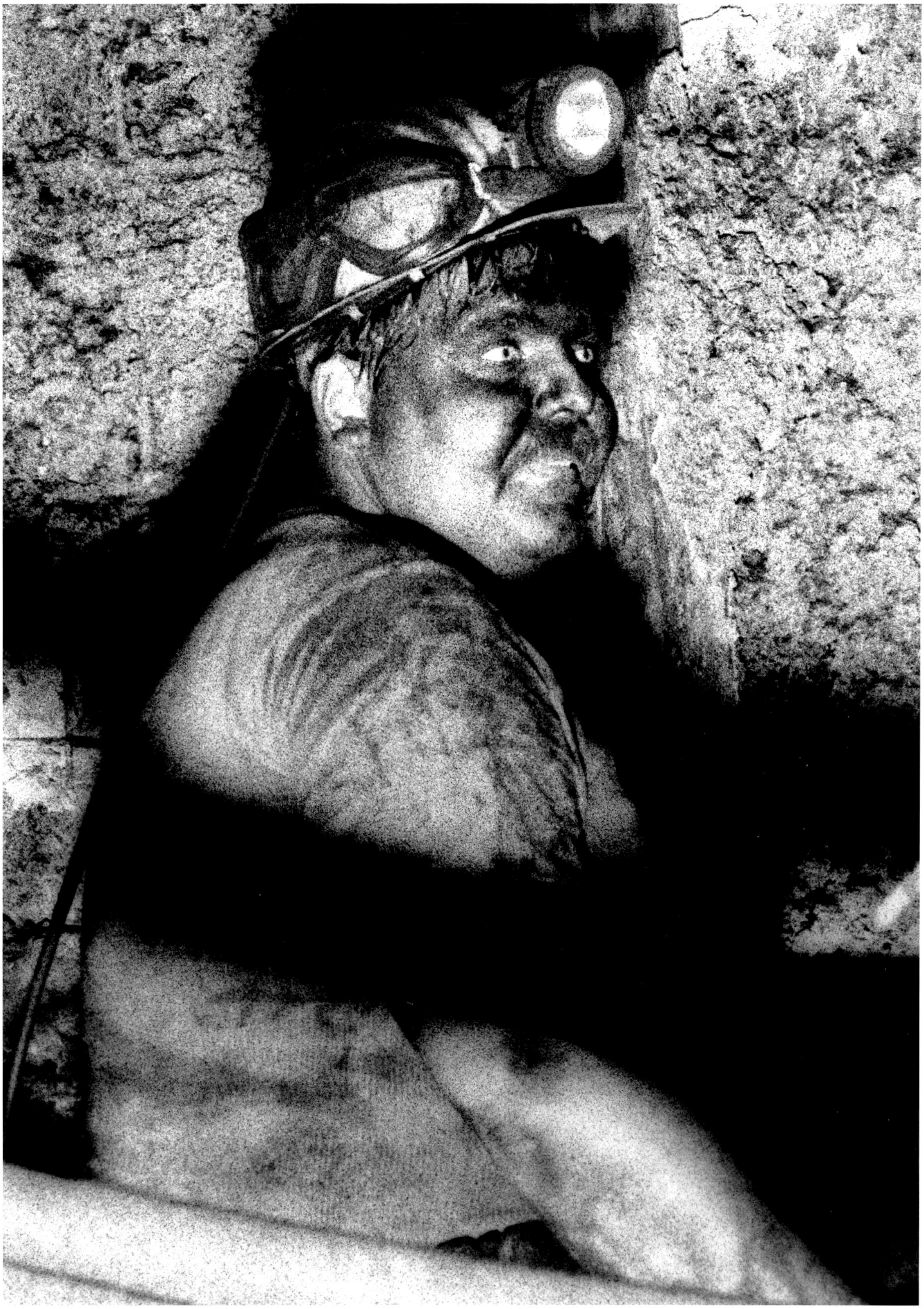

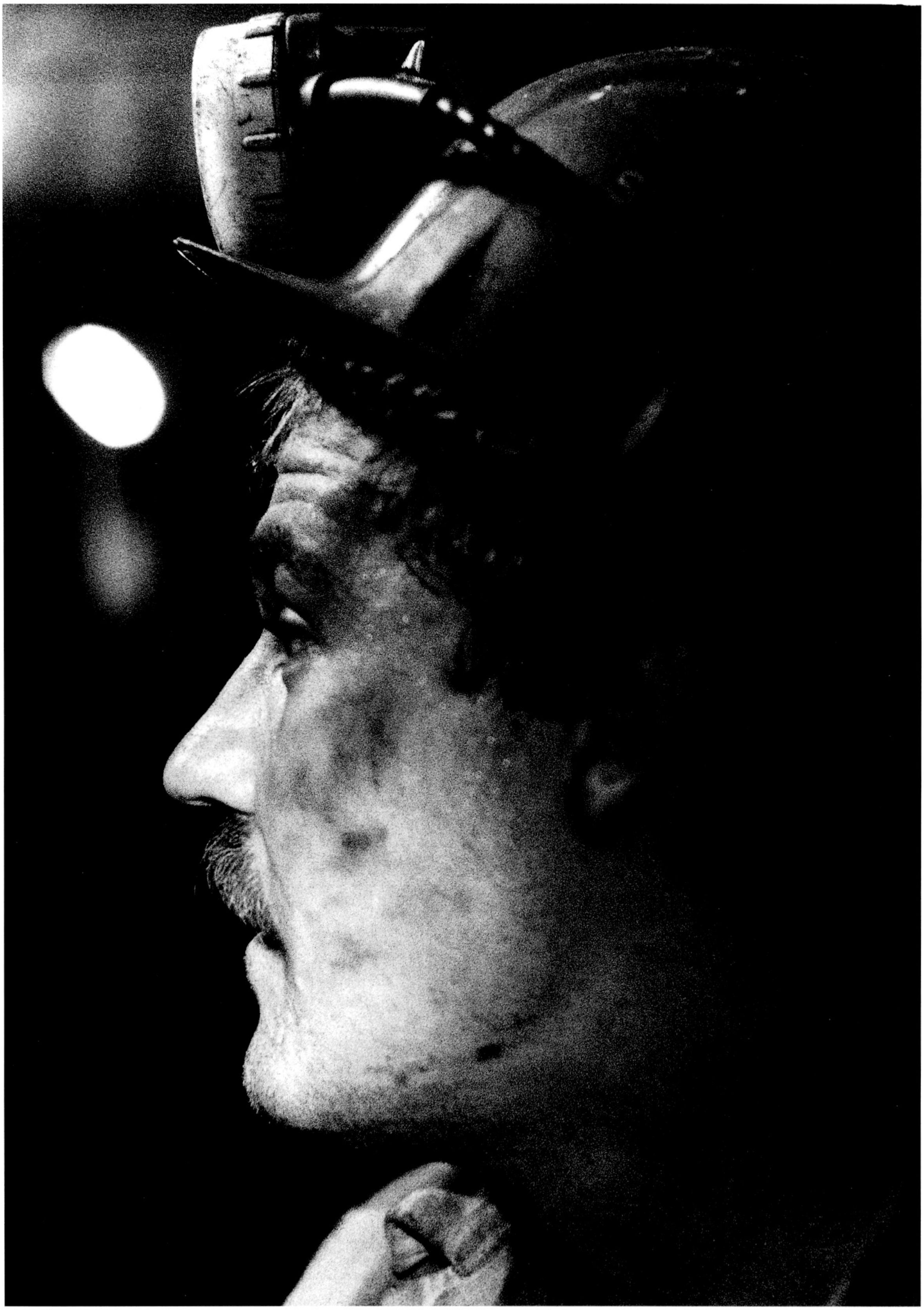

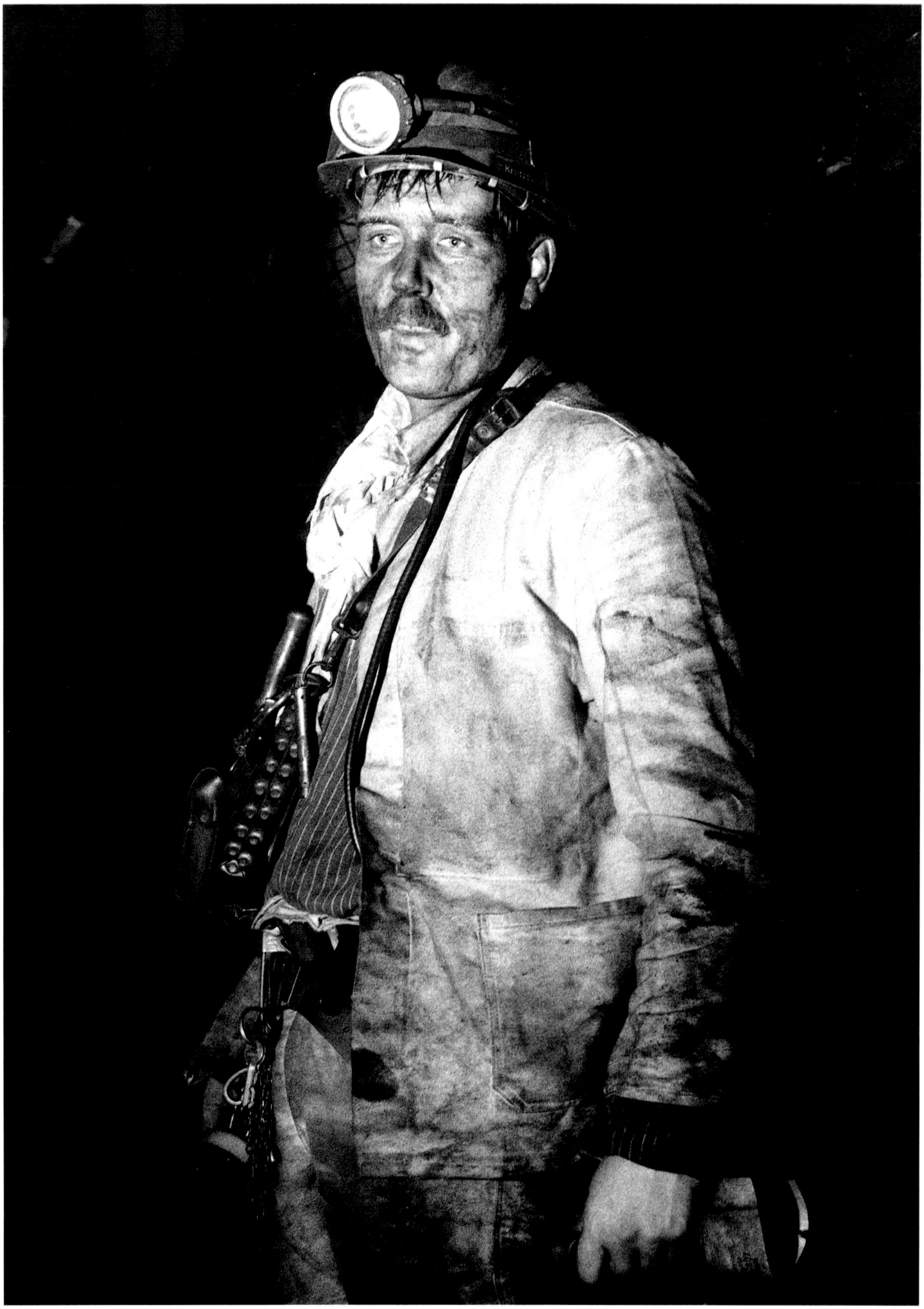

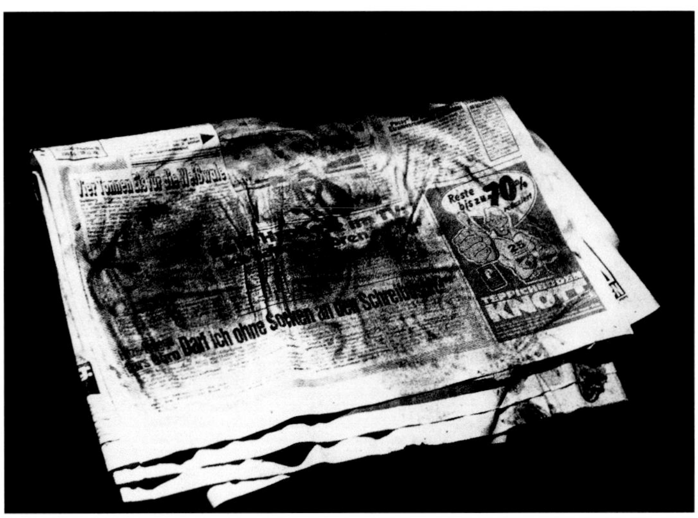

Issue of the “Stadtnachrichten” of 13 July 1994, used as a seat cushion

COAL PRODUCED PER MAN AND SHIFT
IN A LONGWALL ROOM: 33.4567 TONS

Prisen – expression for the practice of sniffing – usually mentholated – tobacco, which gives the miner a short break.

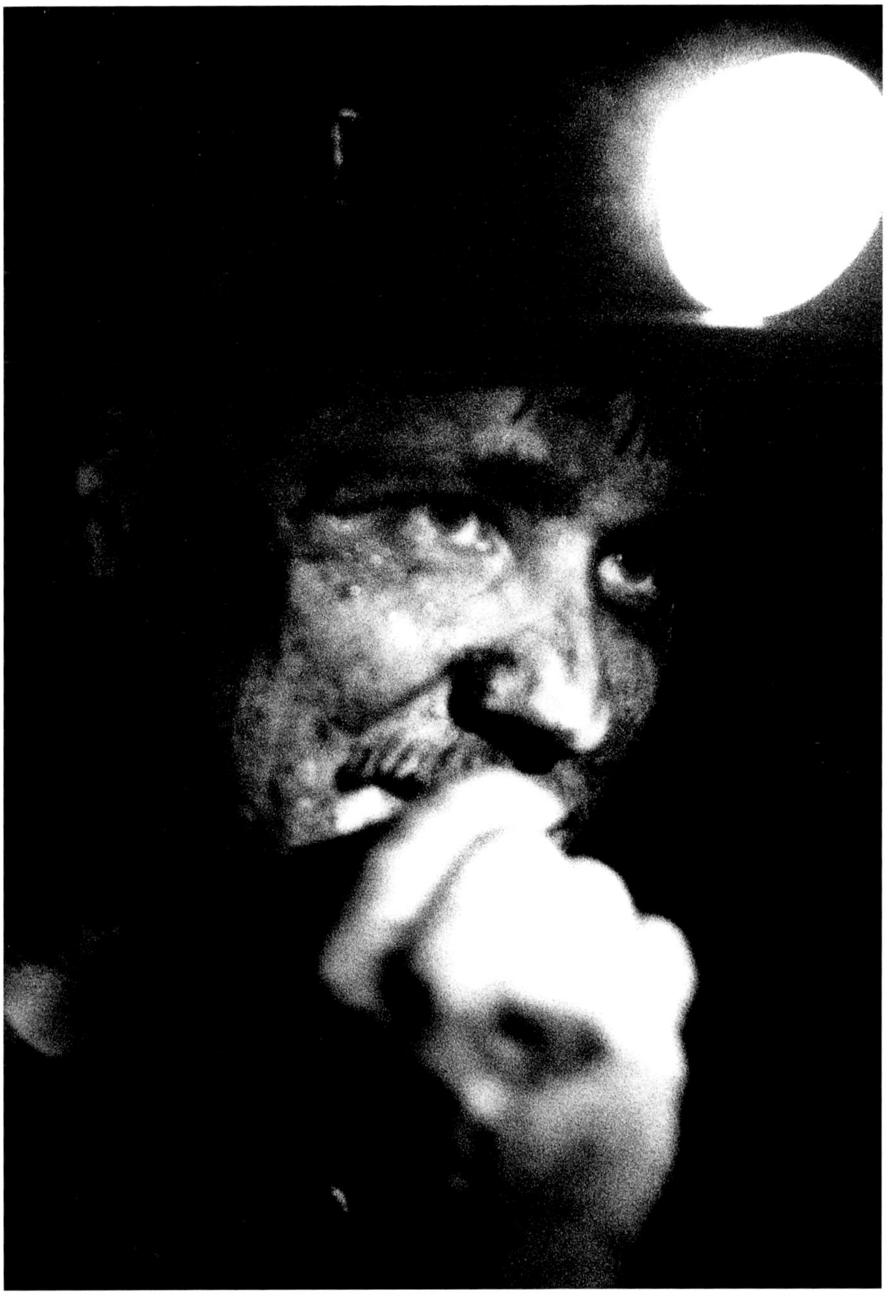

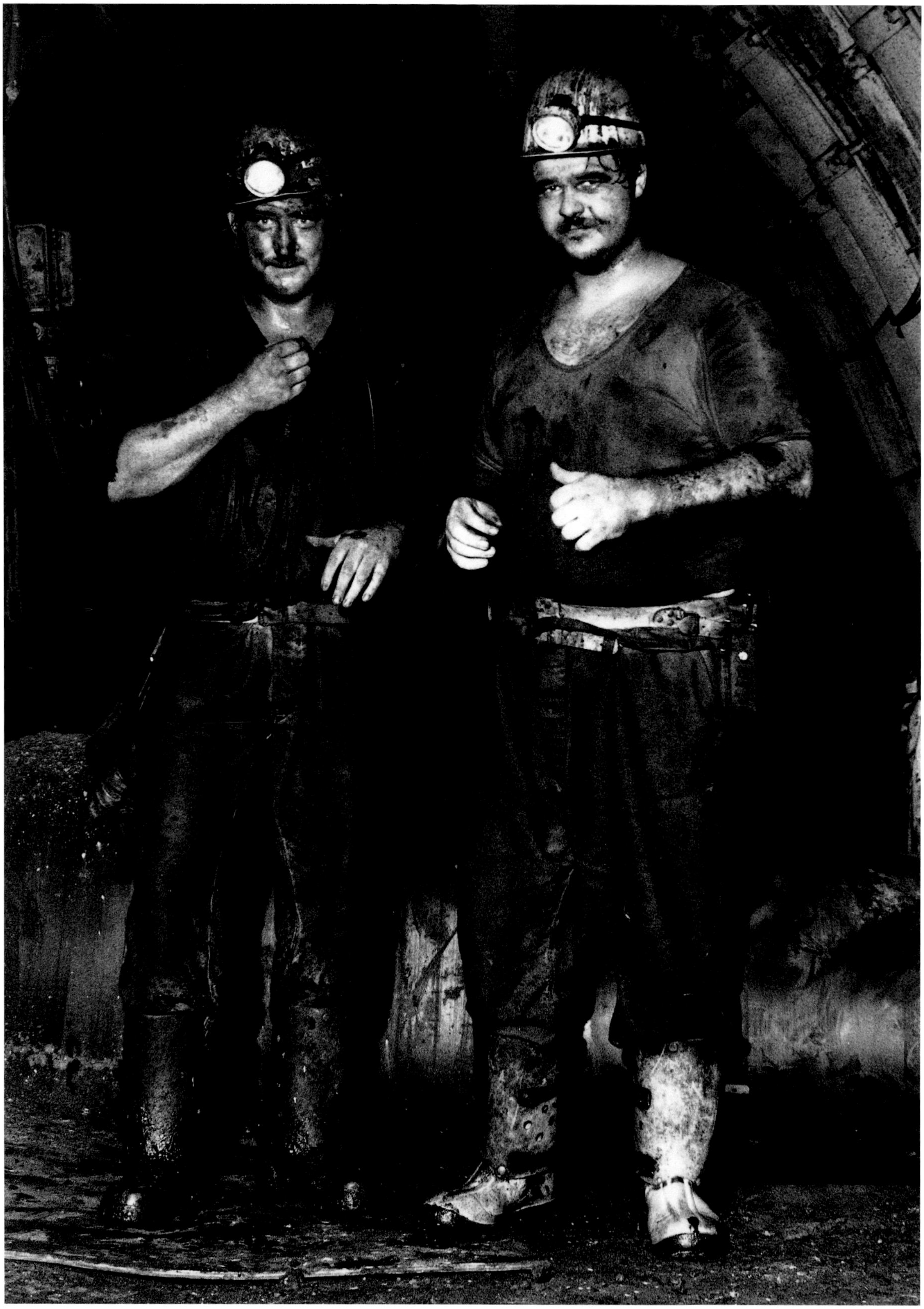

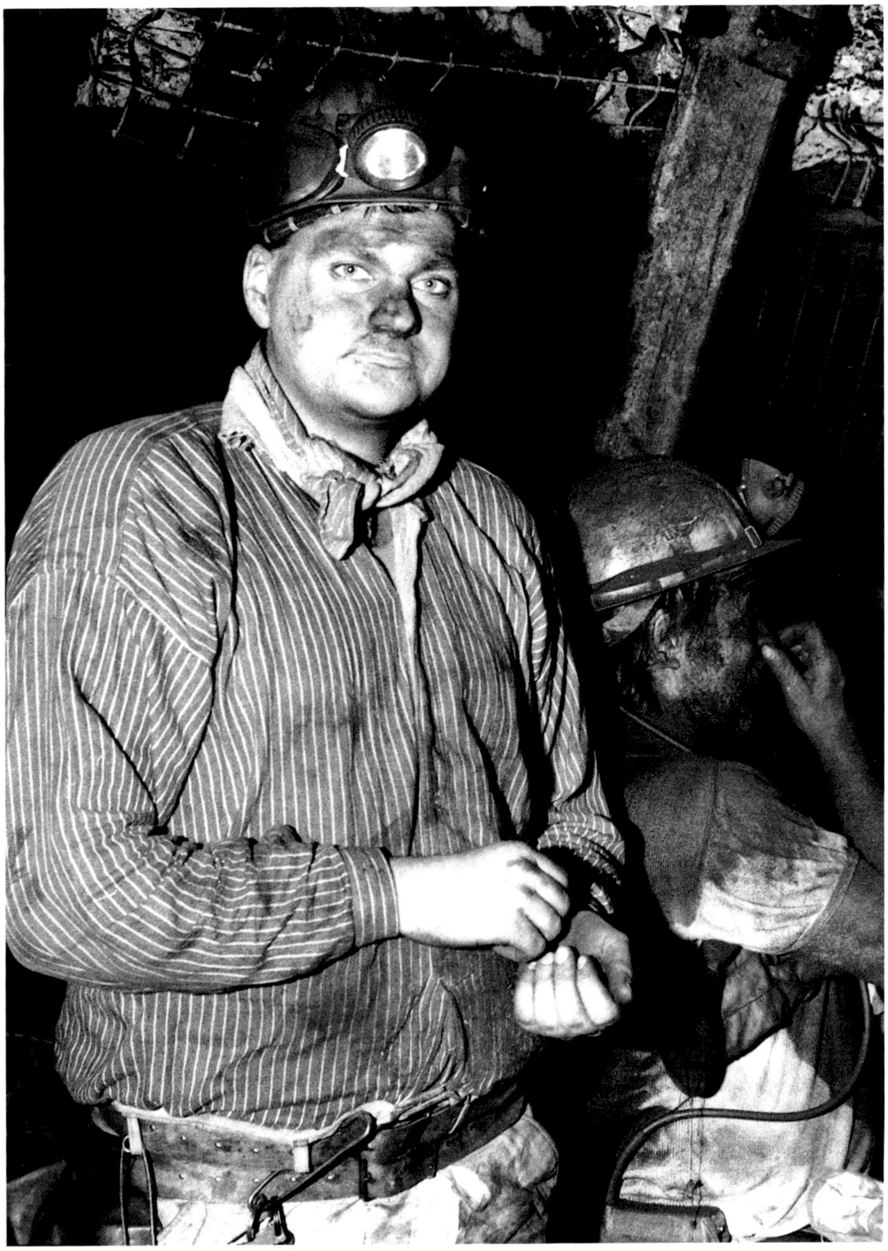

Dubbeln – expression for the break during which miners eat their home-made sandwiches.

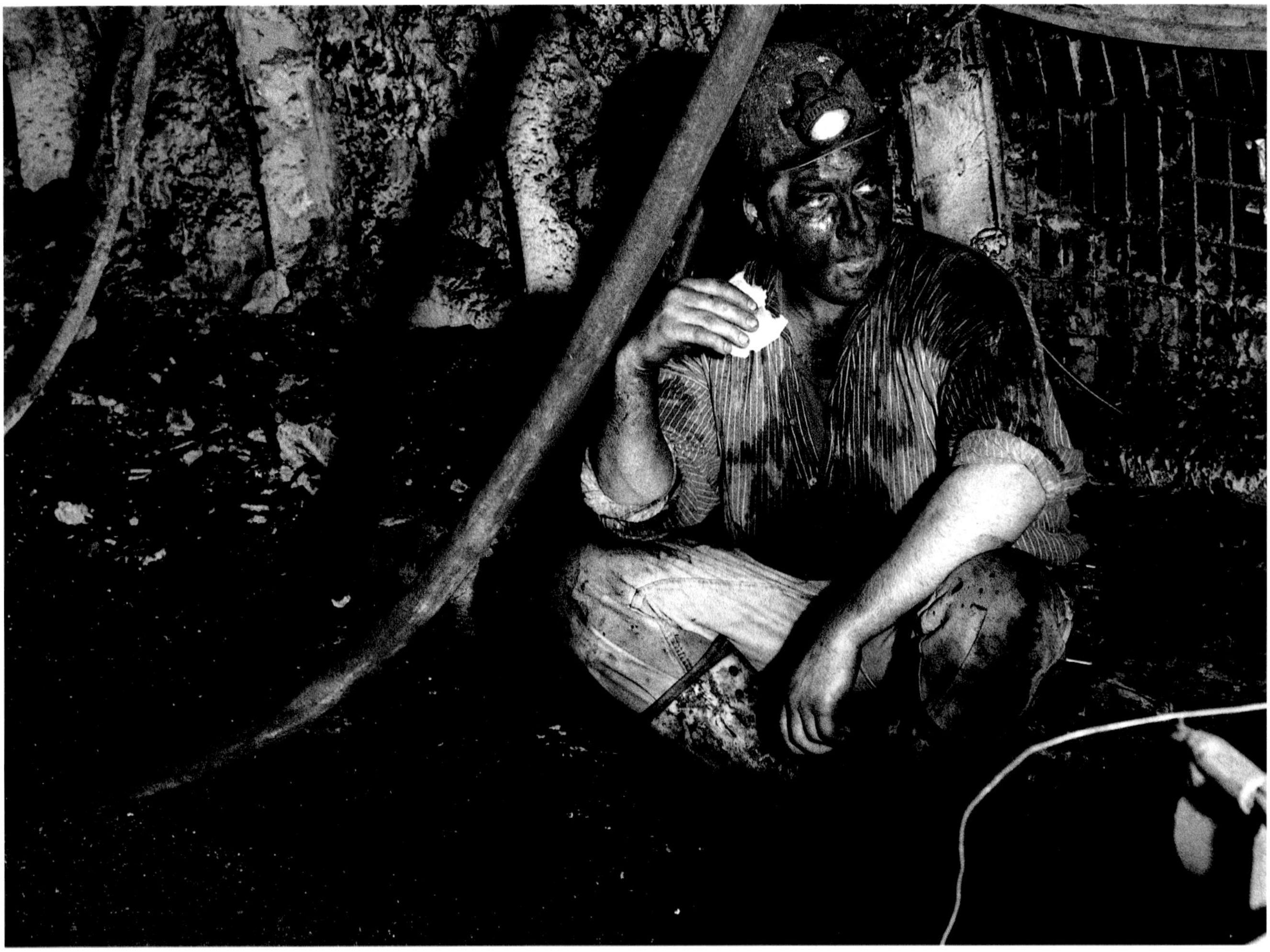

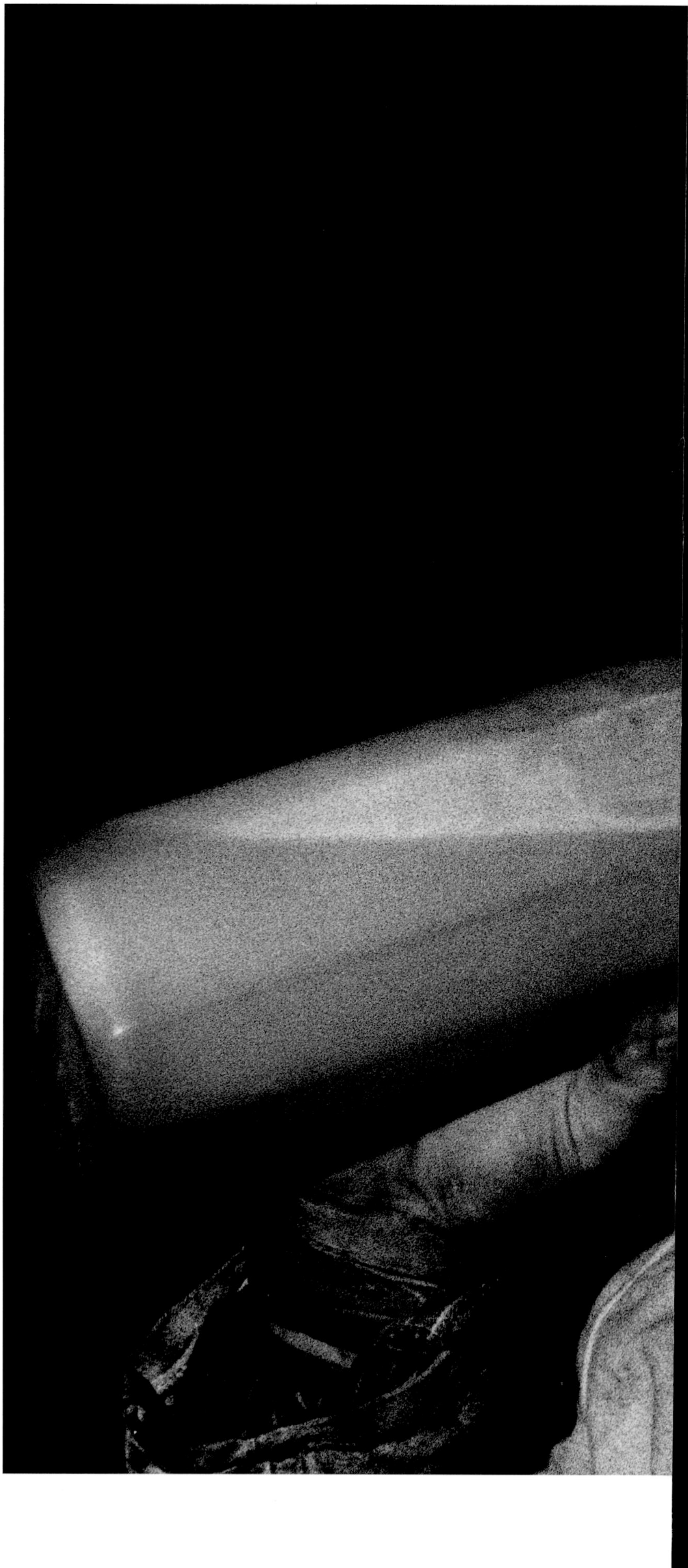

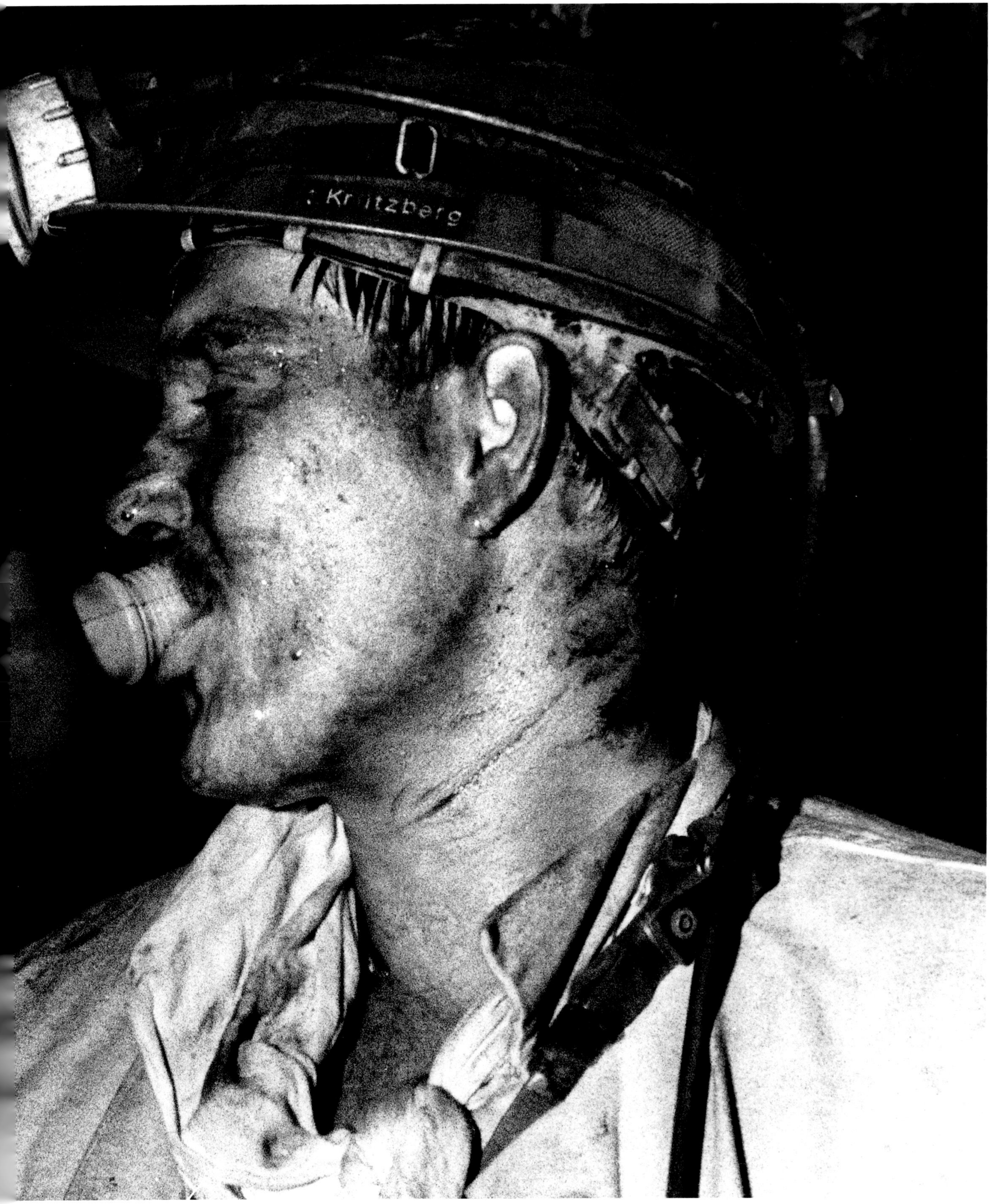

Weatherman – a technician who tests air and climate conditions in the mine-shaft complex (special training required)

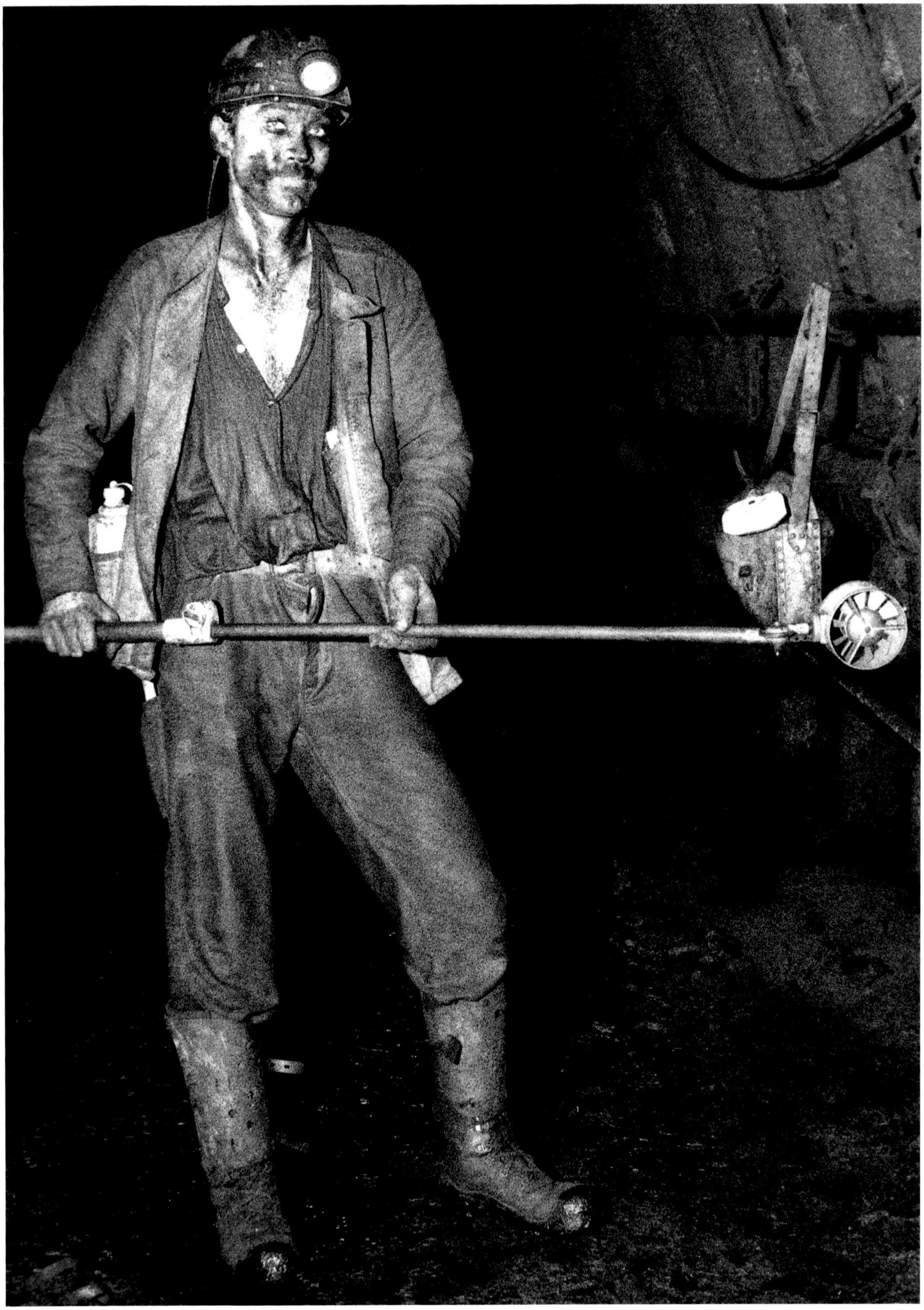

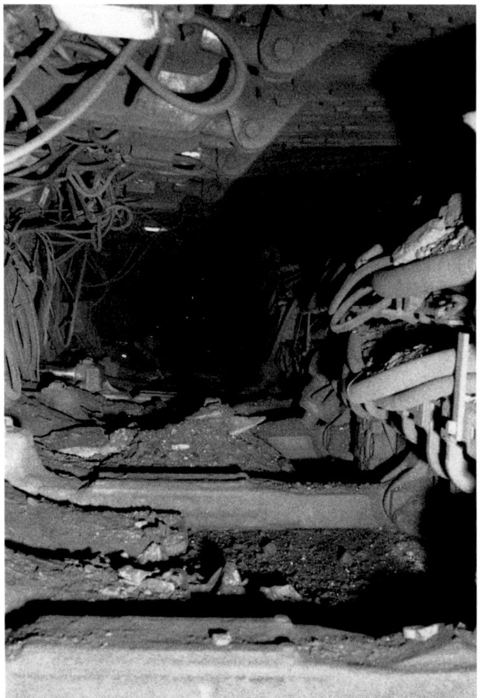

View inside a longwall room

SHAFT LEVEL 551 PRÄSIDENT, KA 38° C

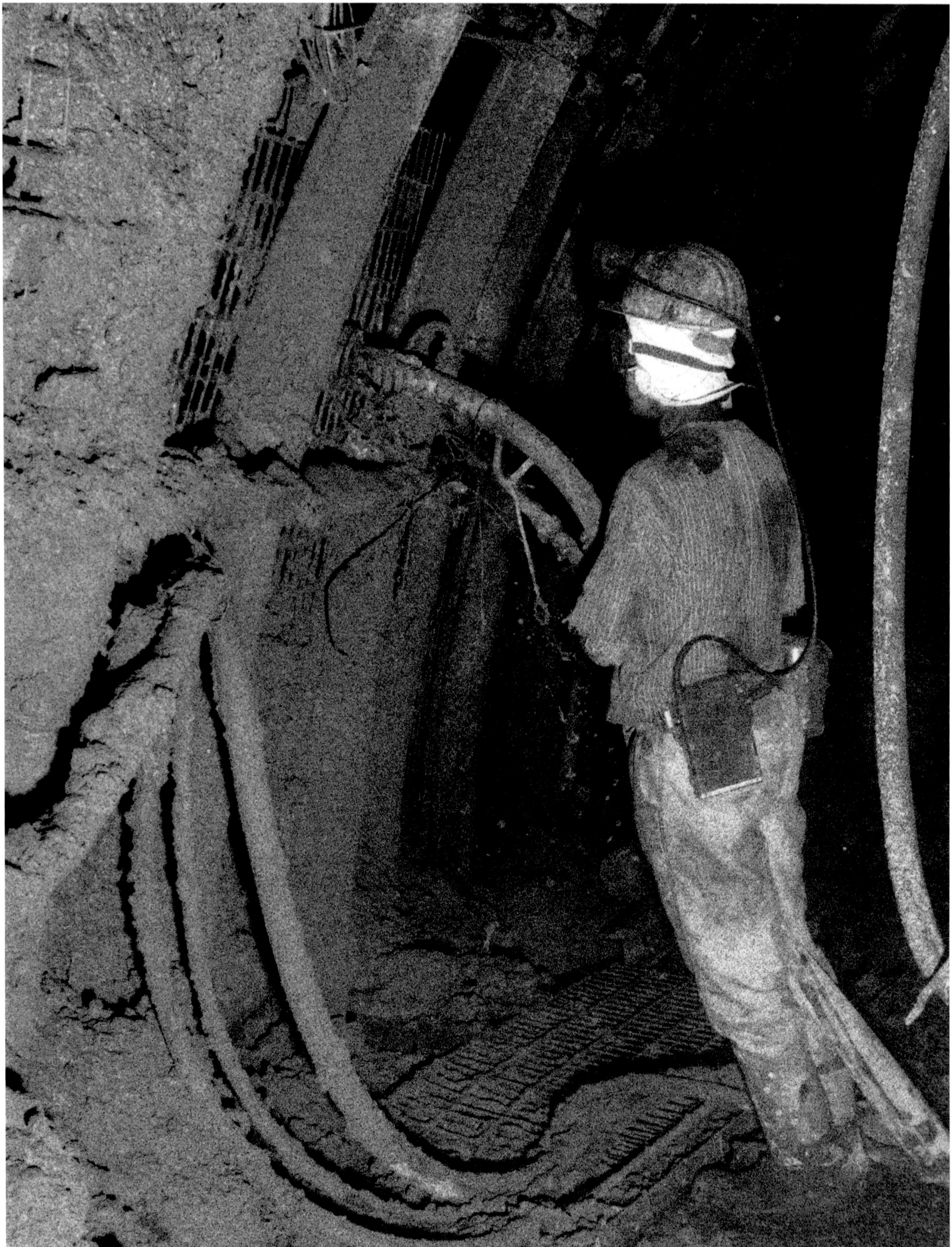

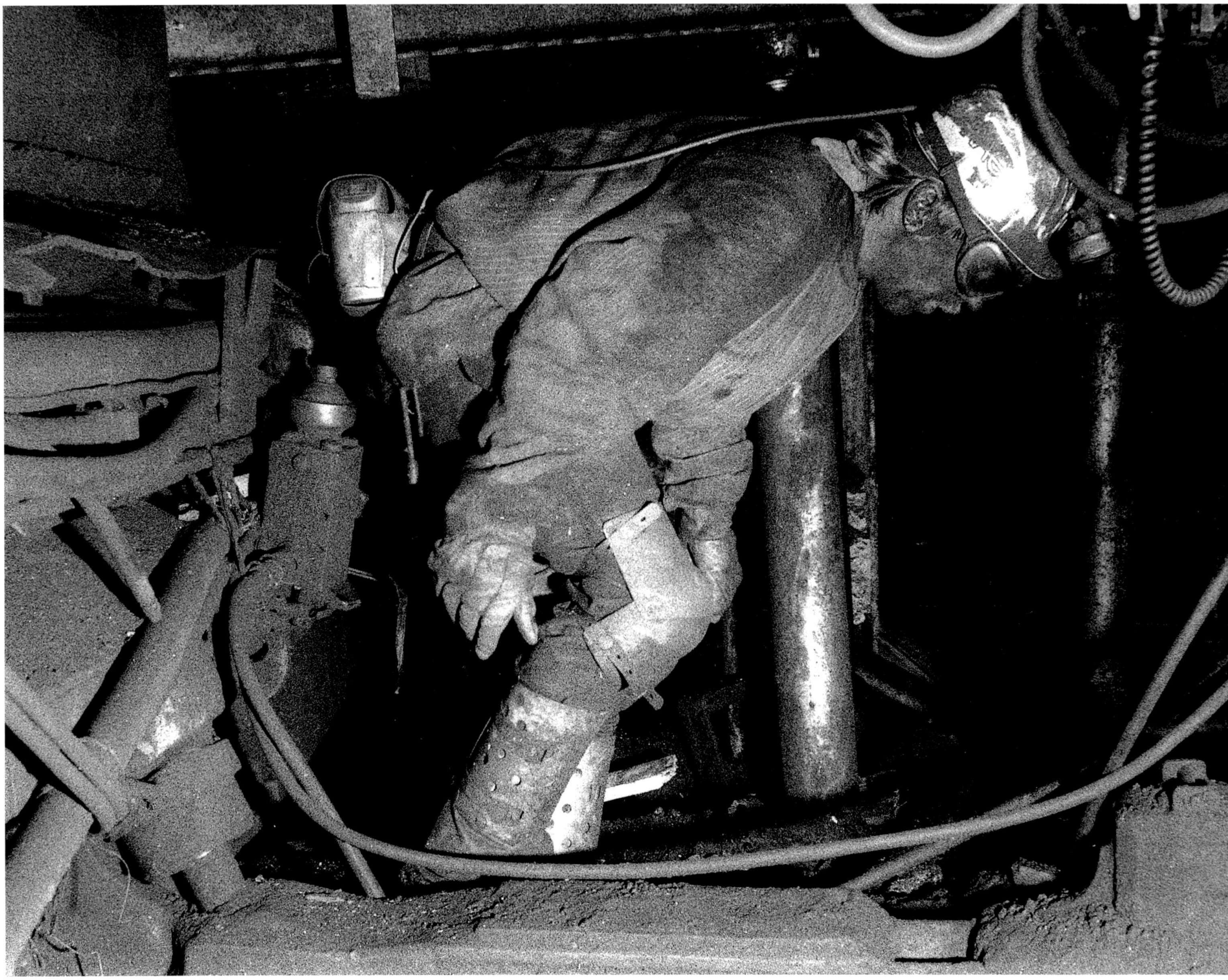

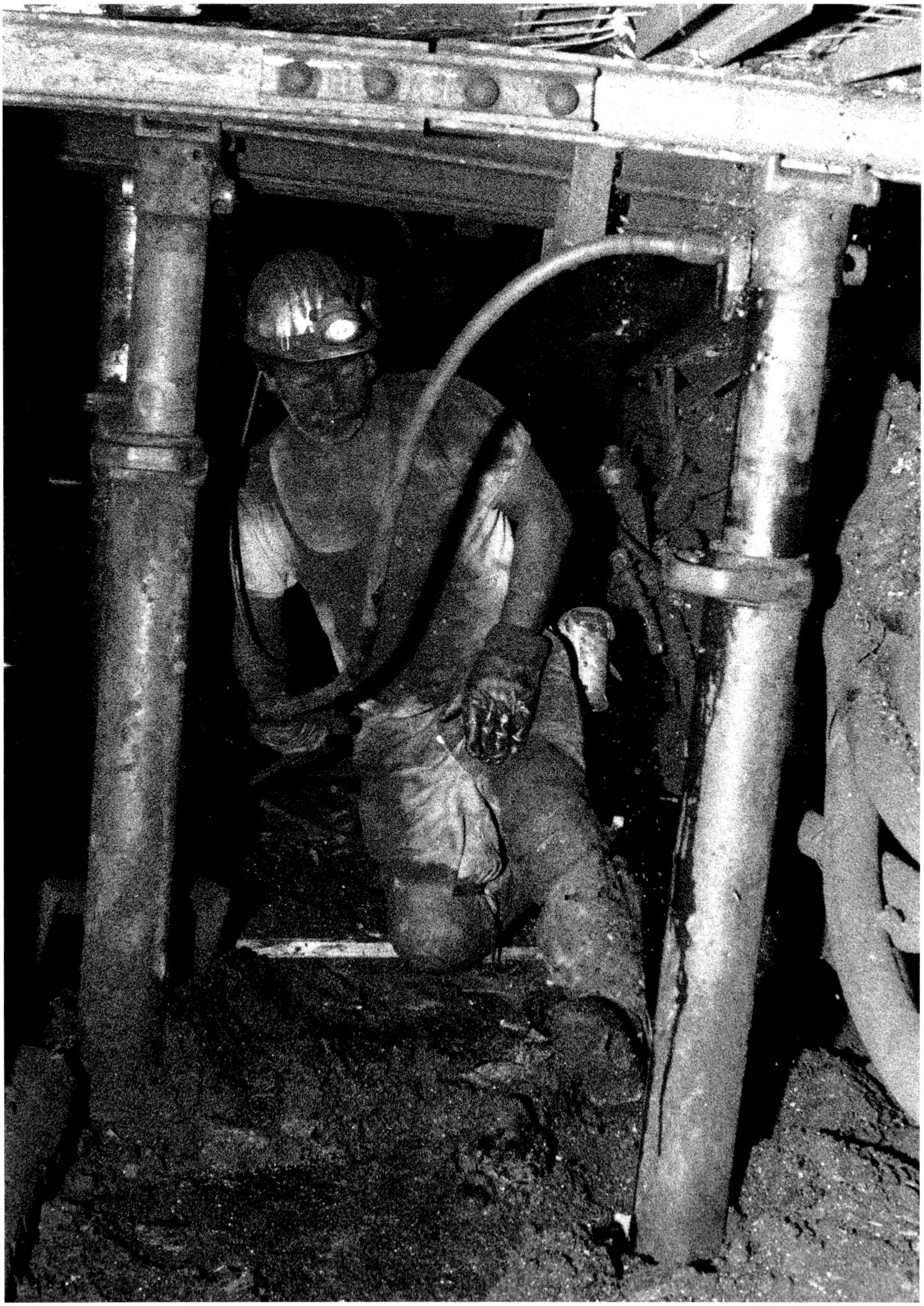

Work-climate regulation – regulation governing on-site working hours, shift times and mandatory breaks as well as other restrictions affecting personnel working underground under precisely defined climatic conditions. Example: 37°– 46° dry temperature, 7.5 hrs. shift time, 4.5 hrs. on site, 30 min. mandatory break time, 15 min. break time based upon a bargaining agreement.

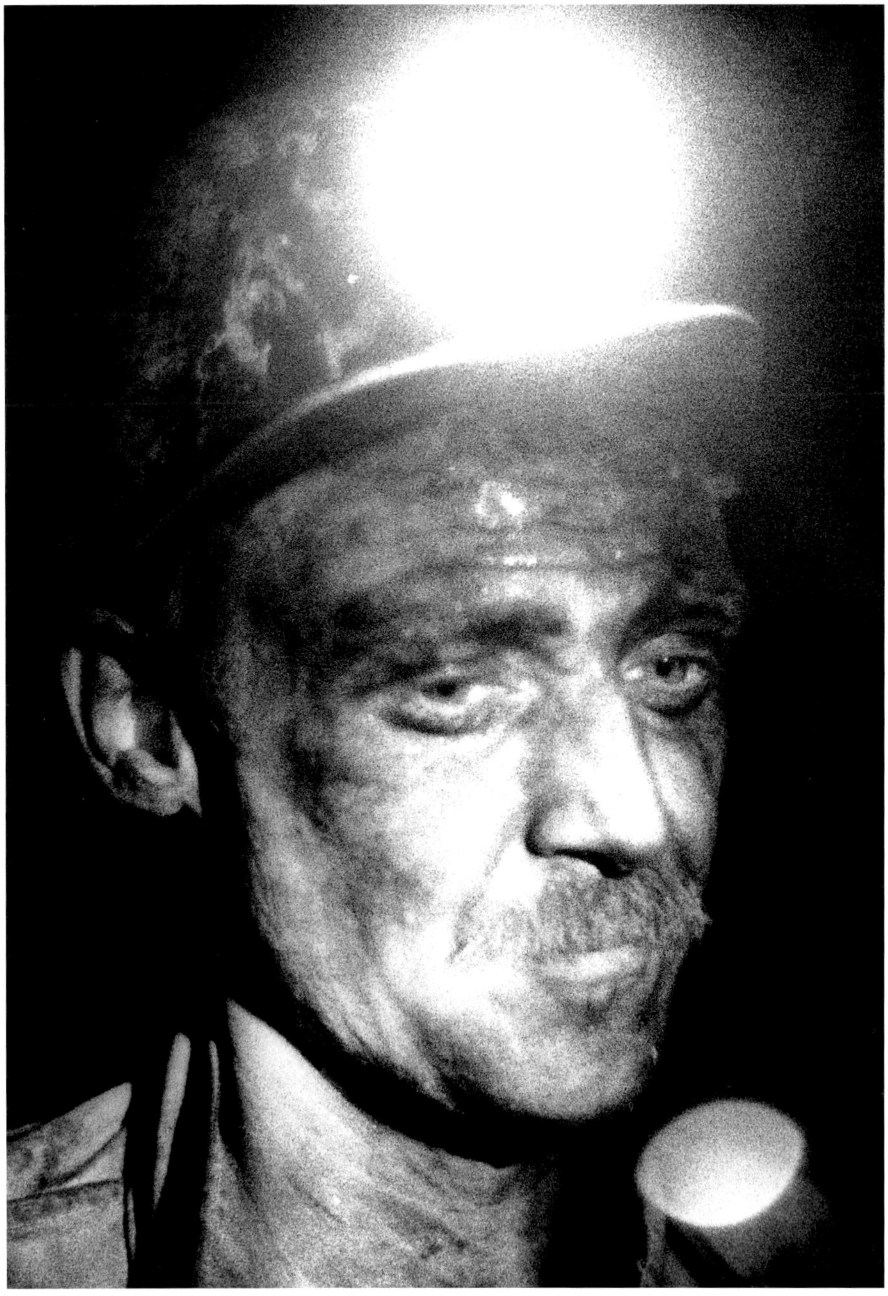

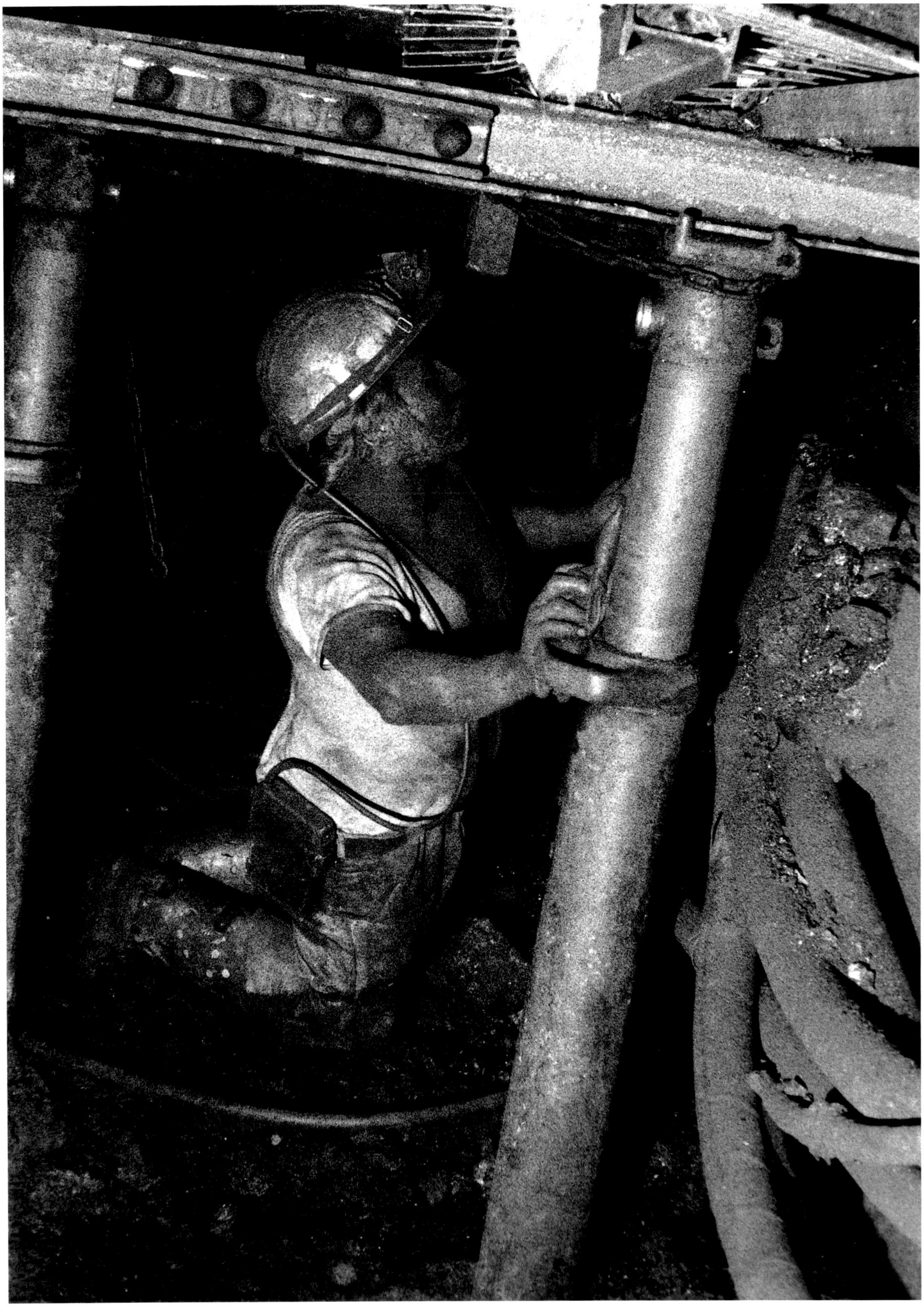

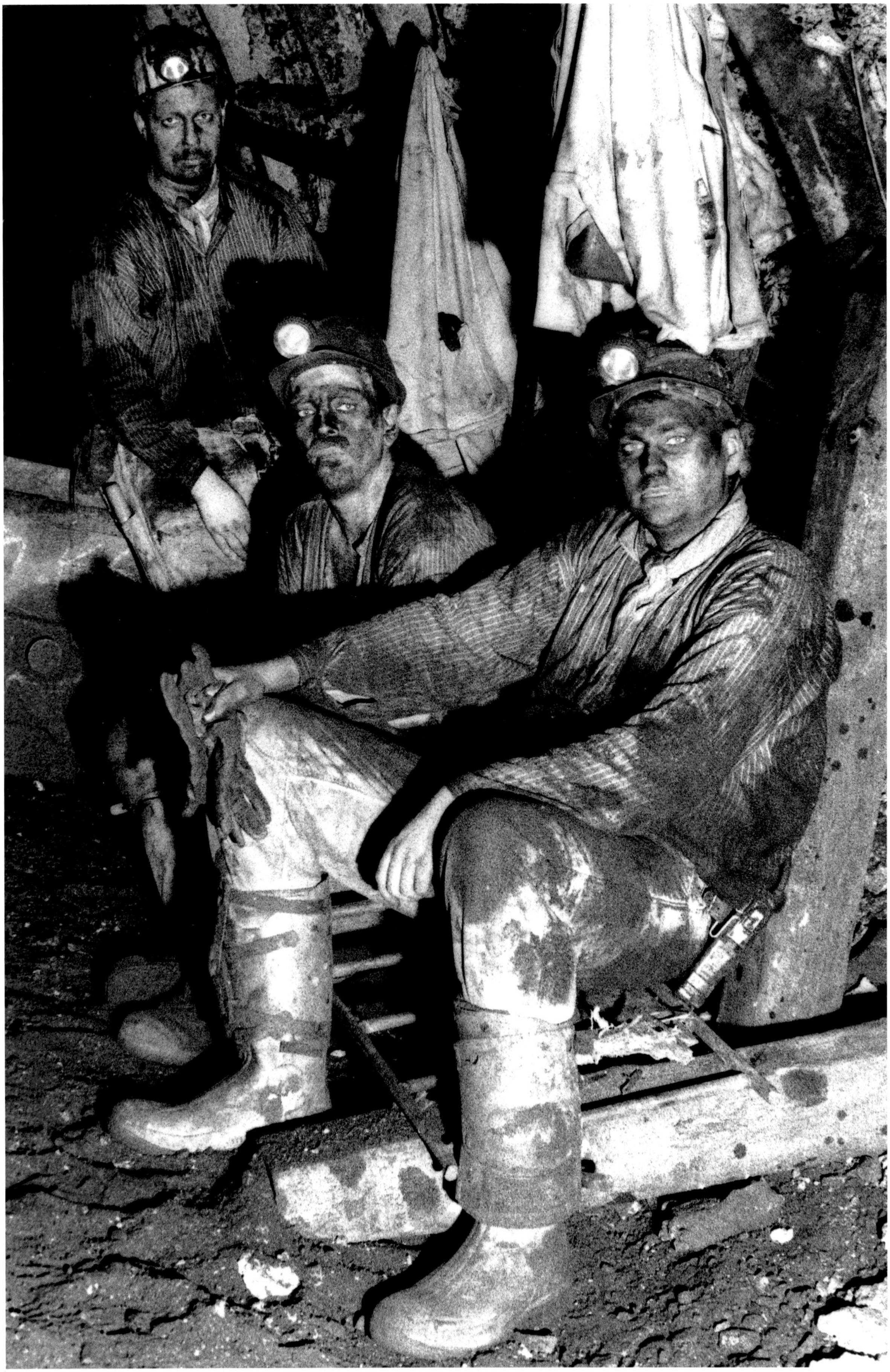

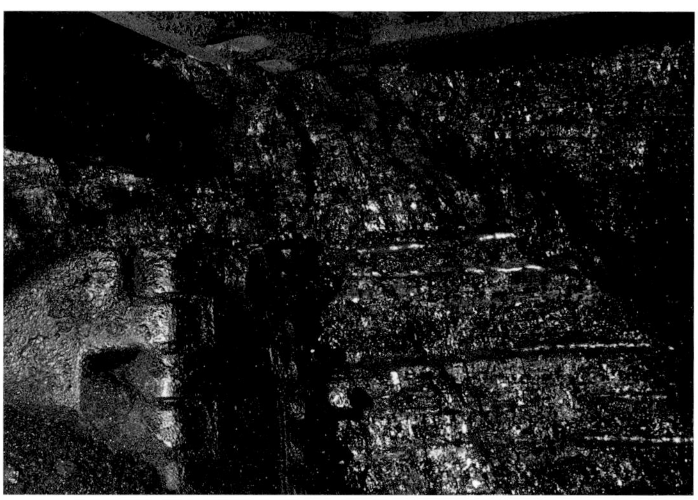

Coal shearer

SHAFT LEVEL 551 PRÄSIDENT, KO 40° C

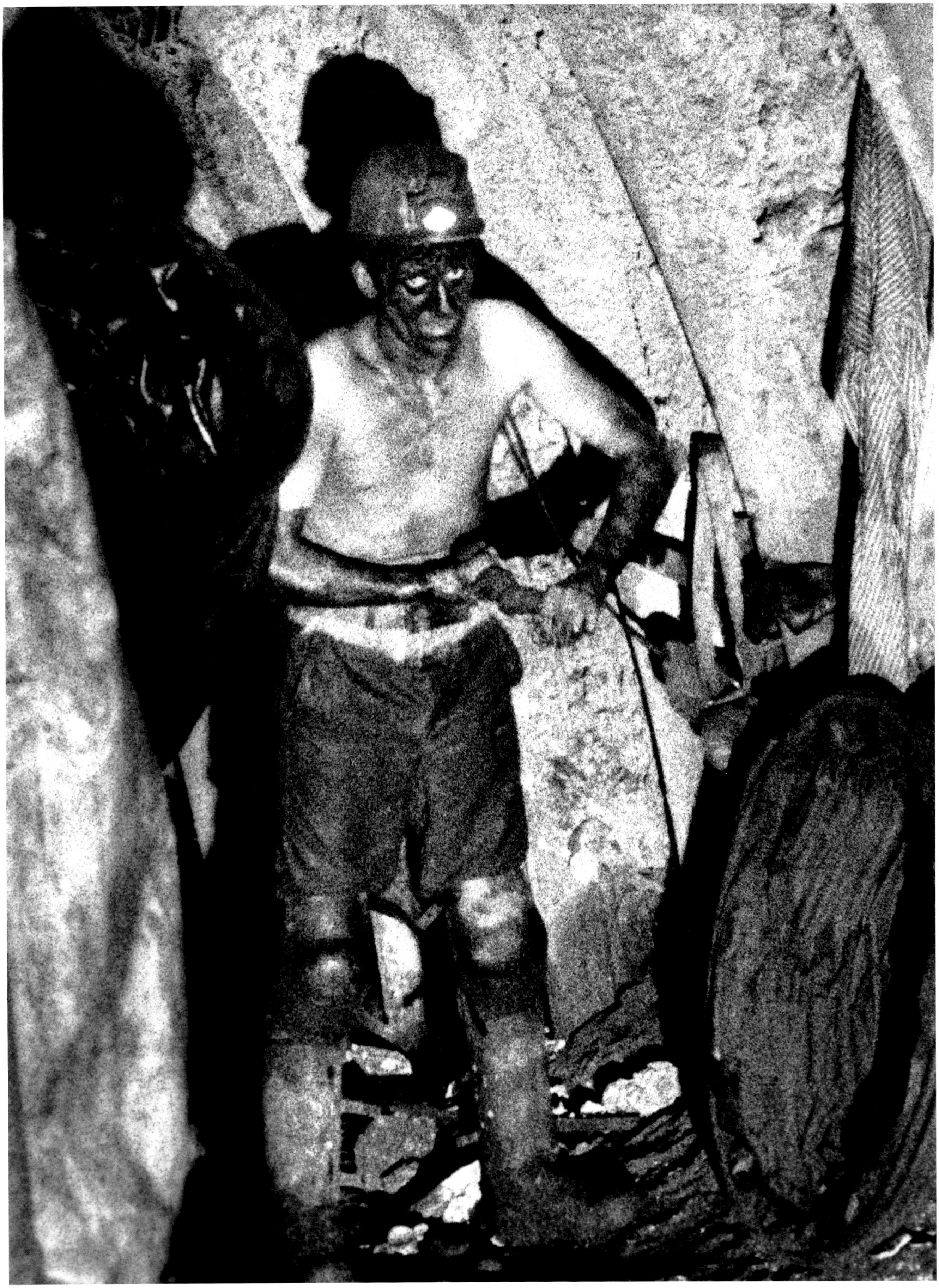

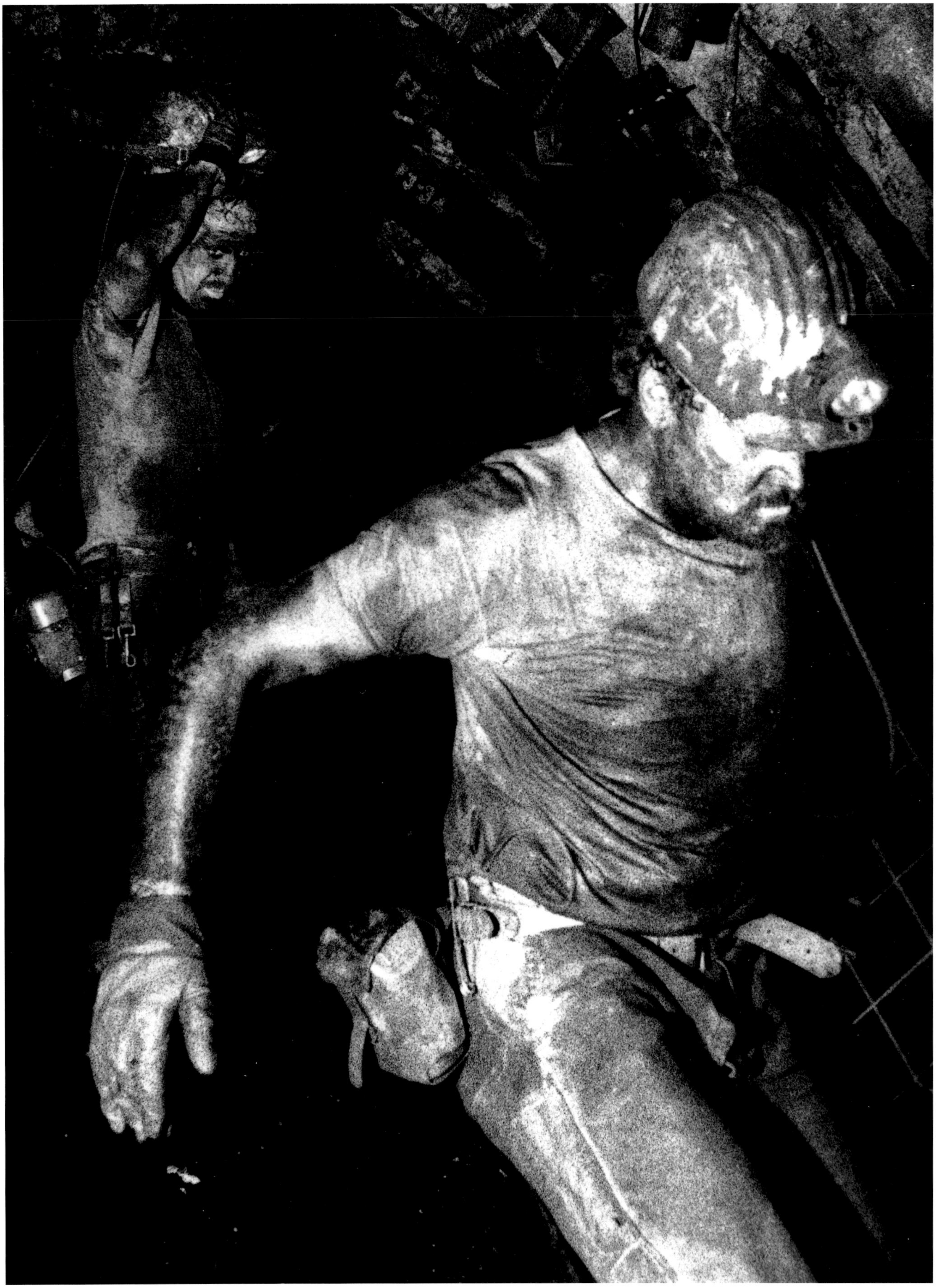

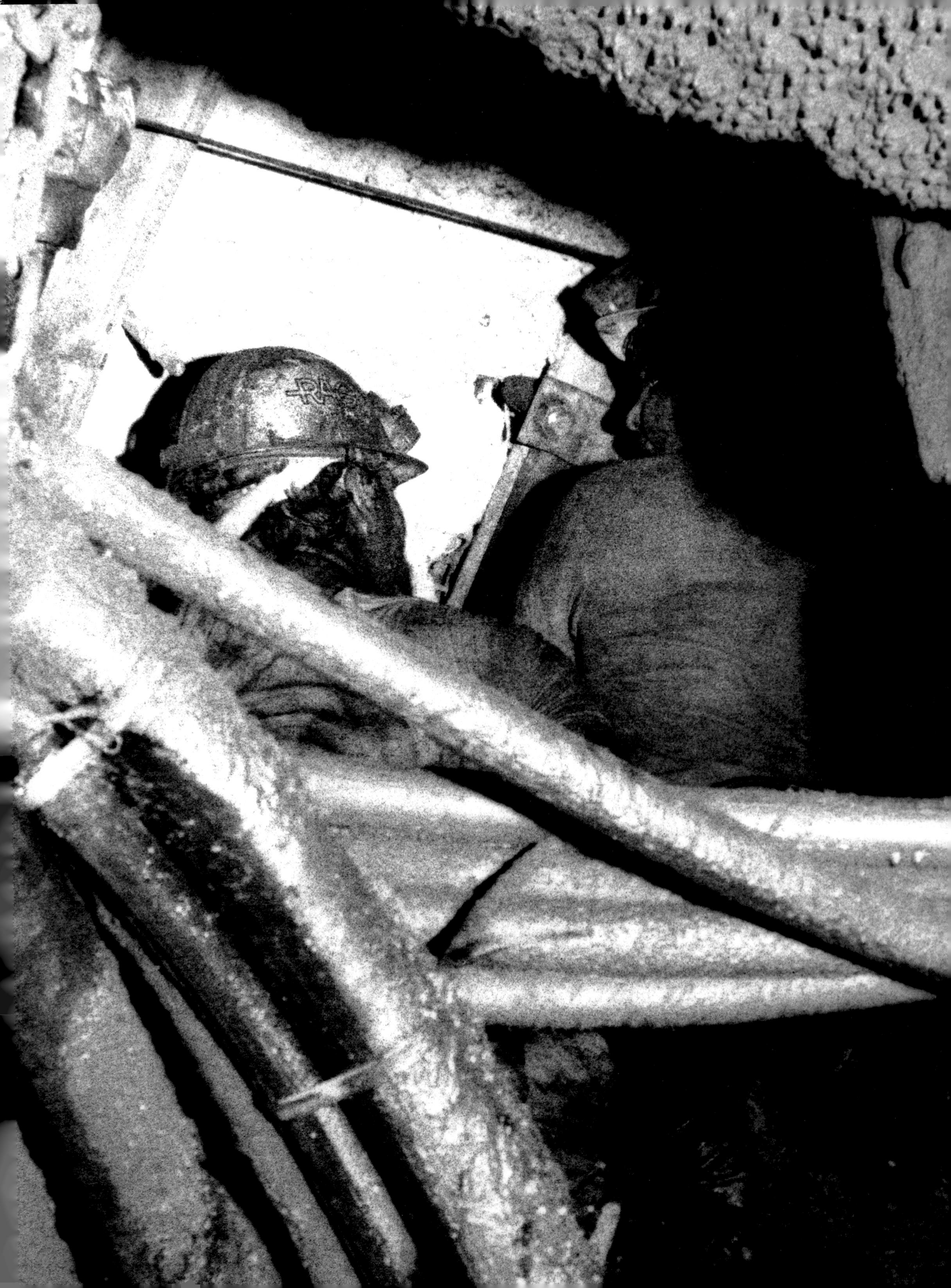

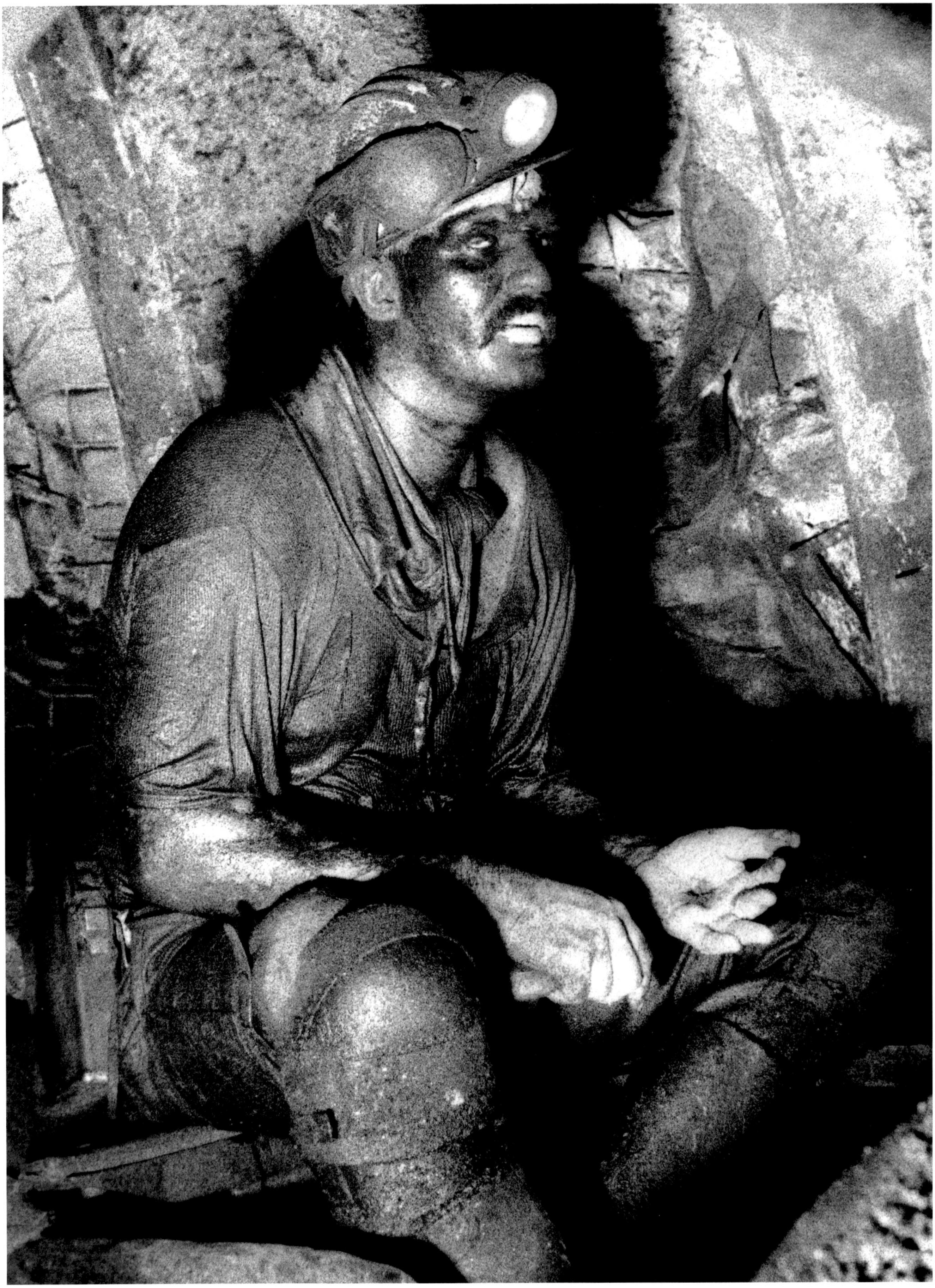

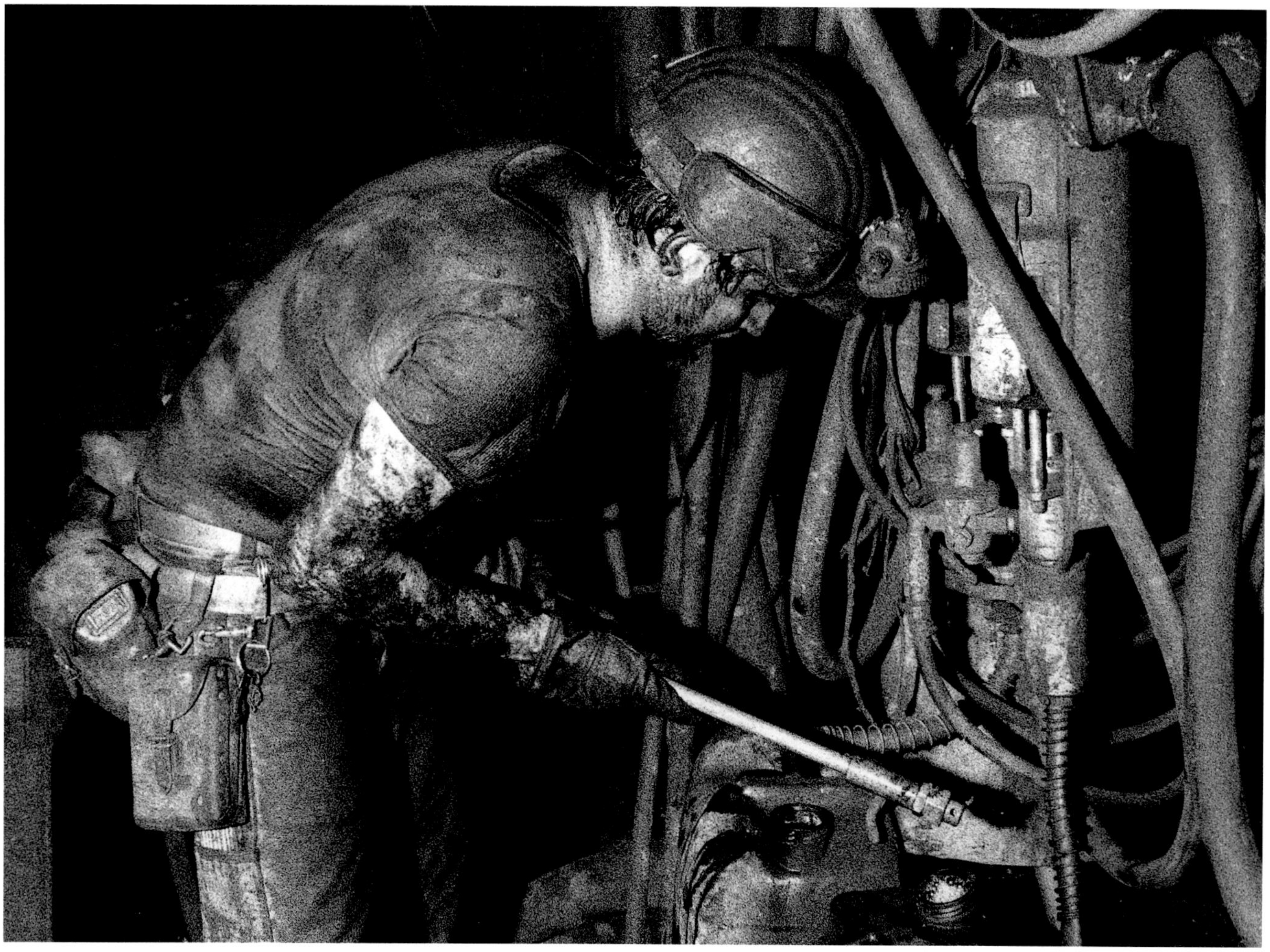

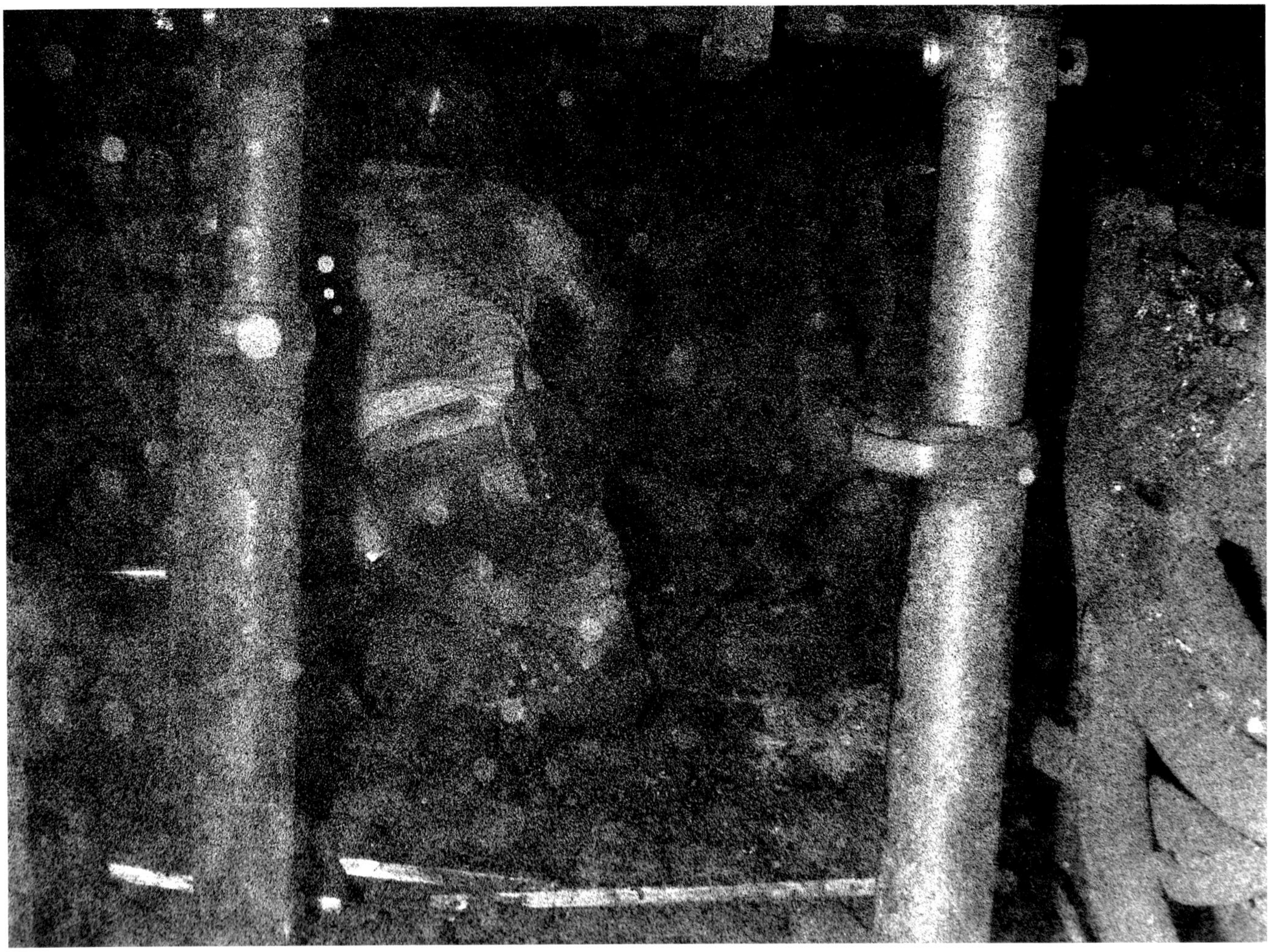

Caving-in a roof section – the removal of limited sections of more or less massive secondary rock from seam roof ends in longwall rooms and stopes when secondary rock is of poor (friable) quality.

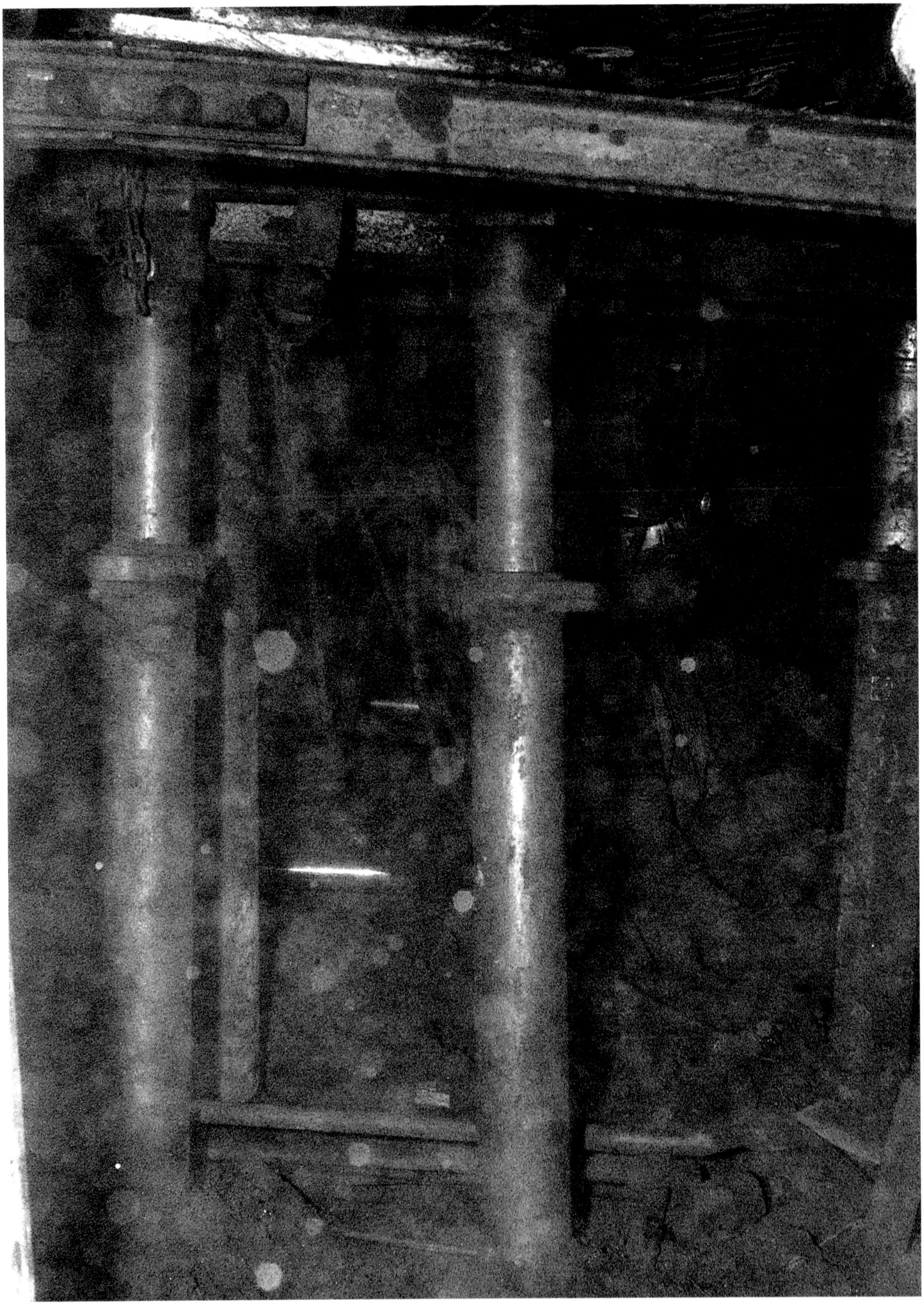

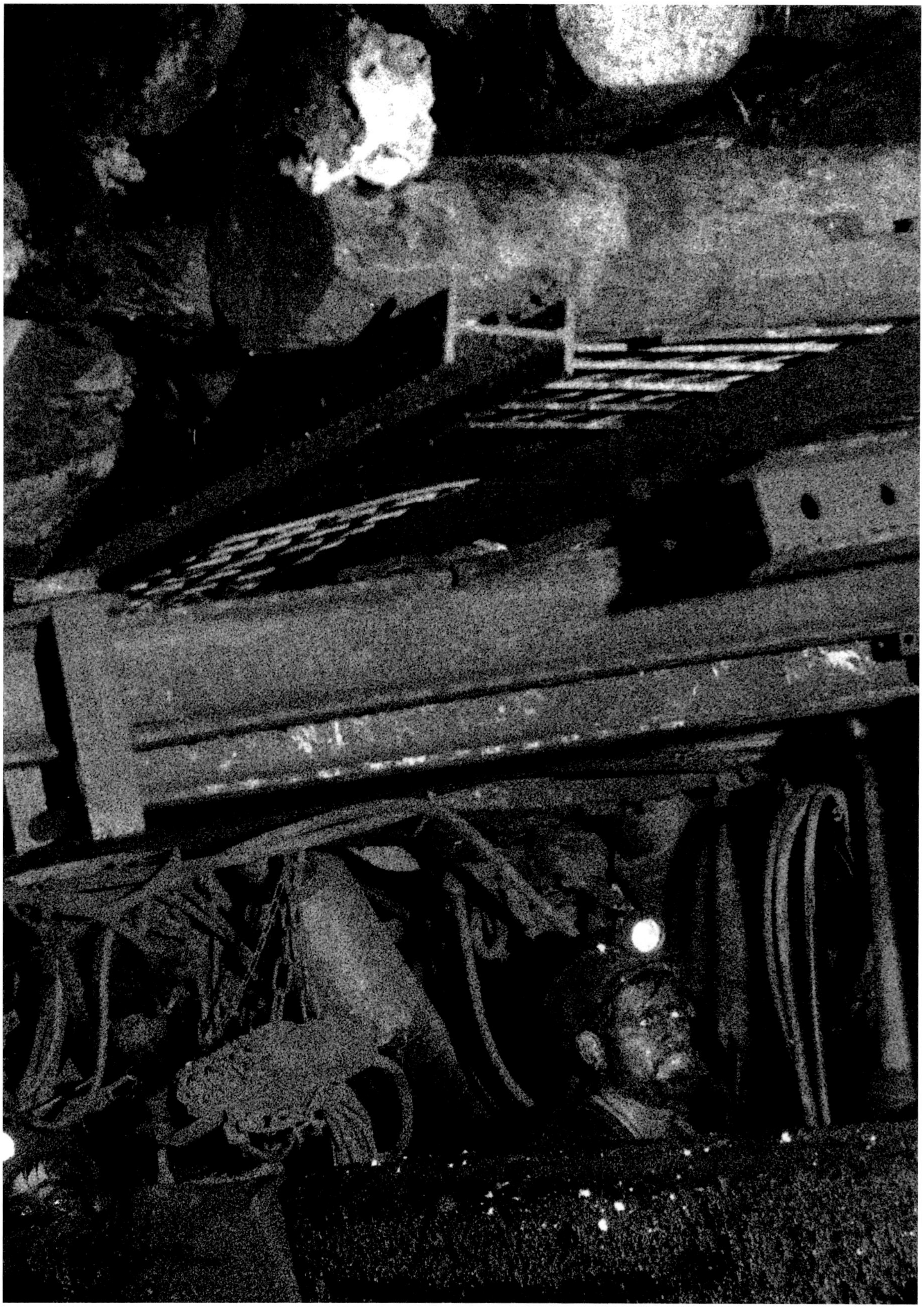

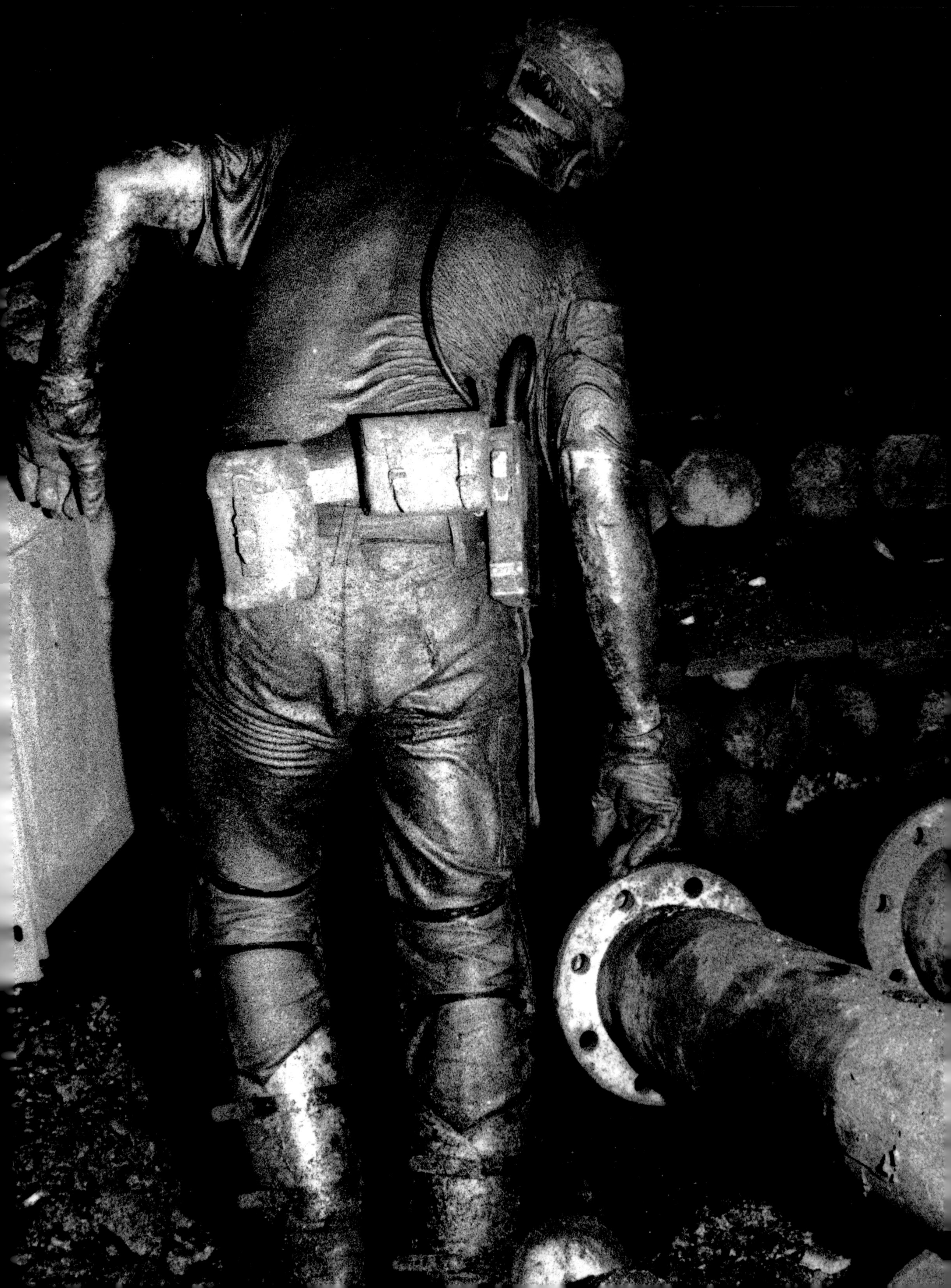

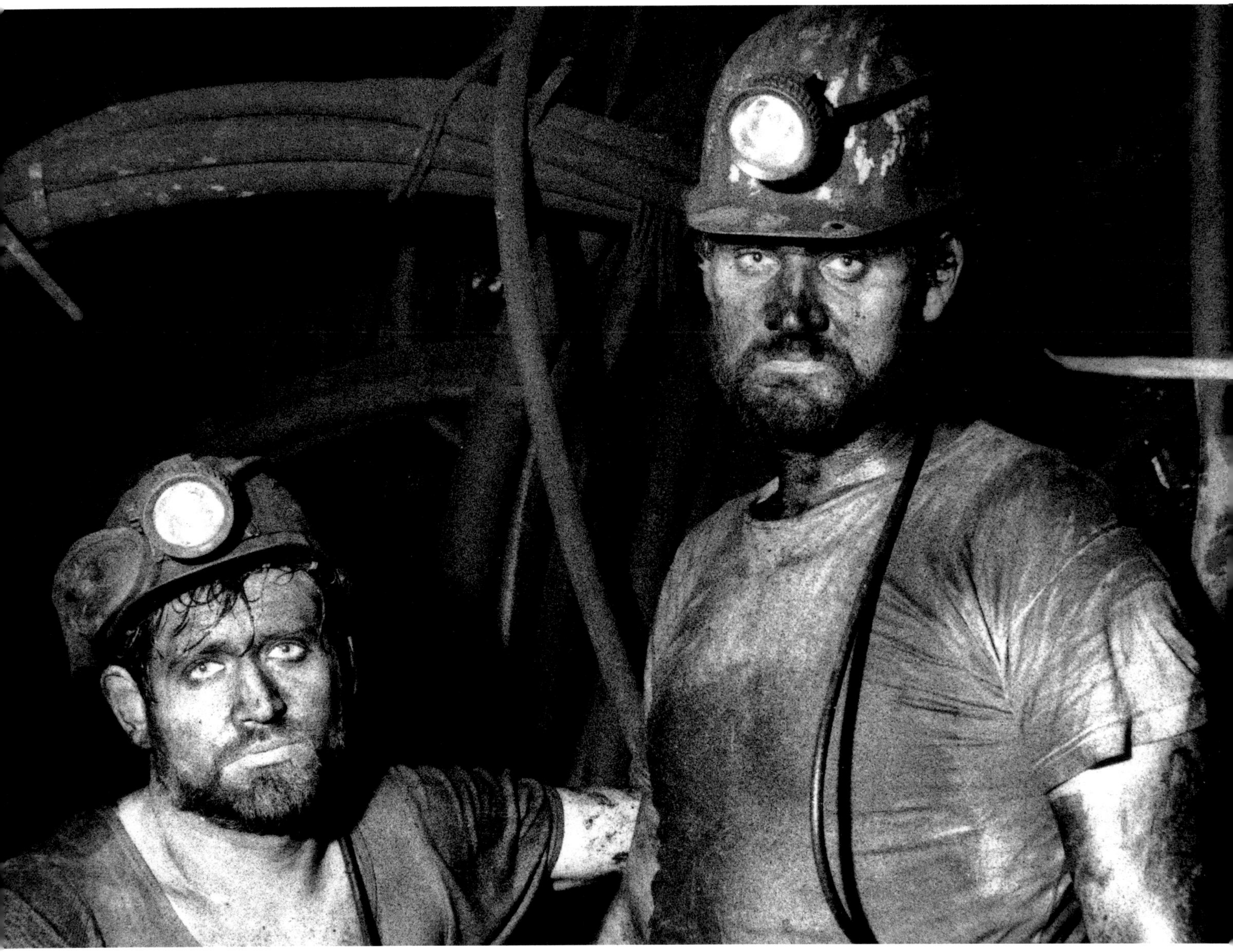

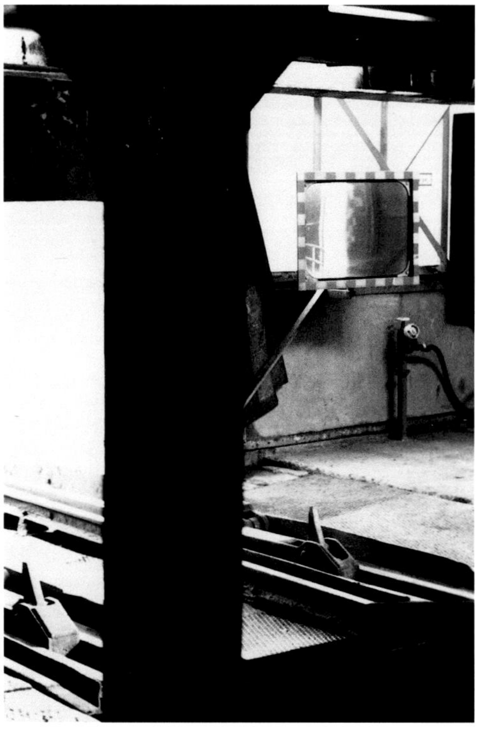

View from inside the miners' cage

„GLÜCK AUF! DU HOLDES SONNENLICHT,
SEI INNIG MIR GEGRÜSST!
DER ACHTET DEINE STRAHLEN NICHT,
DER TÄGLICH SIE GENIESST."

(Vers from a popular miners' song)

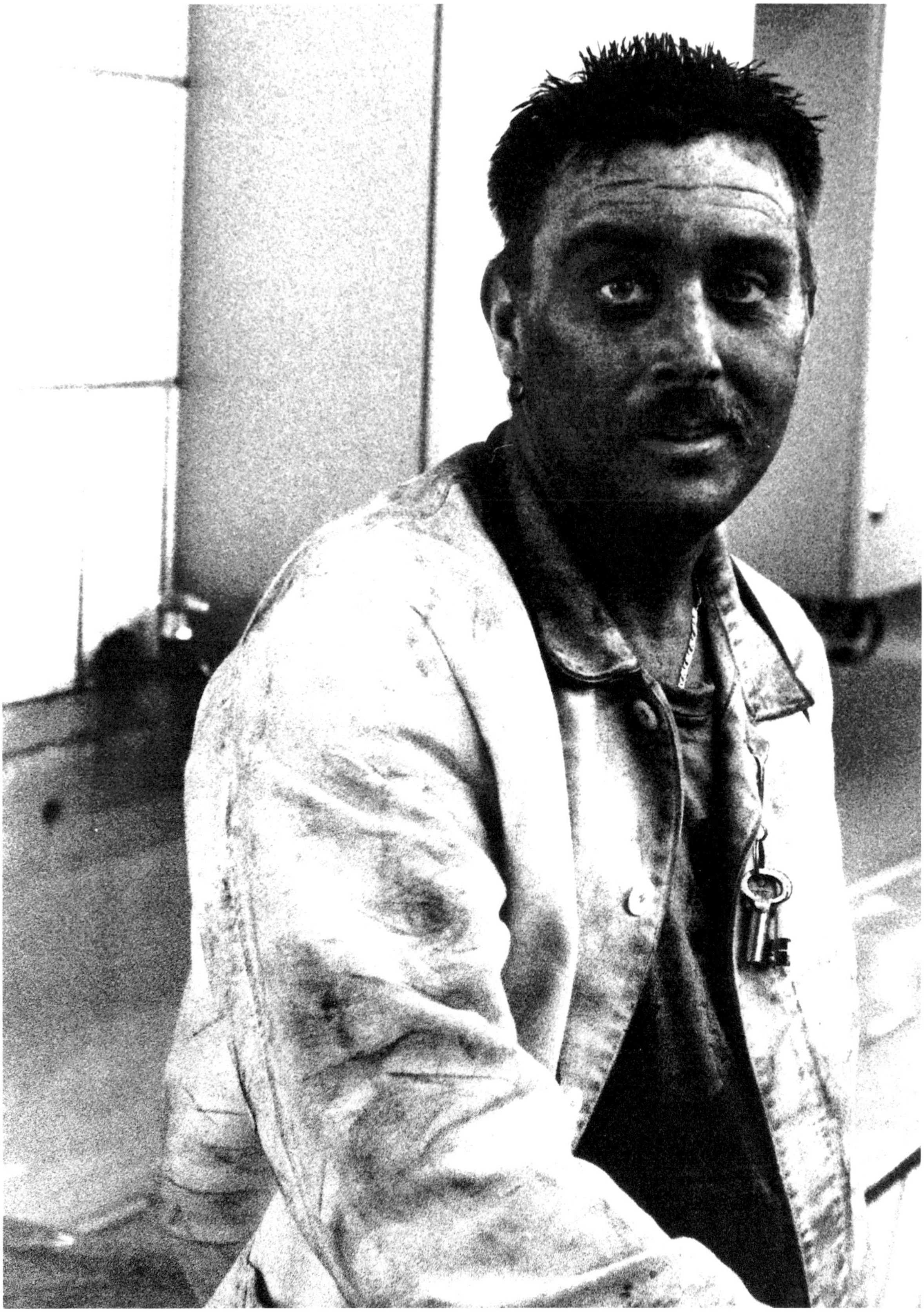

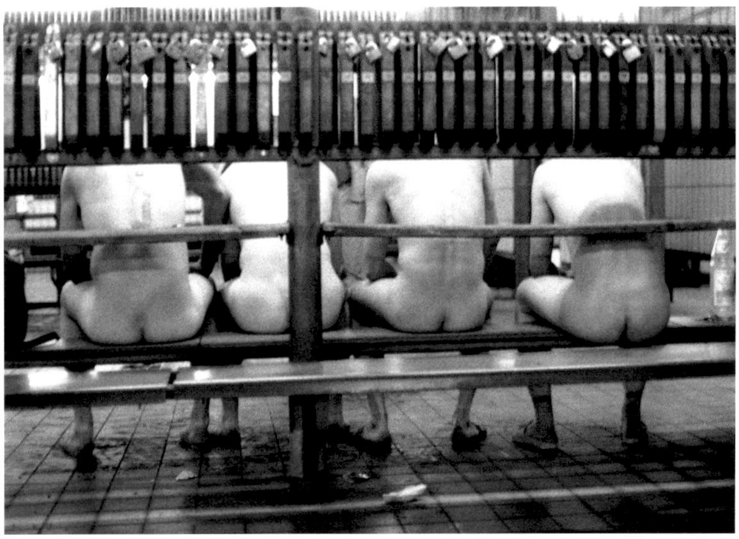

The noise comes nearer. It is like the sound of a freight train racing over rail joints on uneven ground – with heavy links of chain crashing against exposed edges. The air thickens with dust; the clamor is deafening. You feel the ground trembling beneath your feet; the earth shakes. Then the shearer has passed. Particles of coal dust dance in the light of my miner's lamp.
I couldn't see the machine, because it races past this point behind a steel wall. The temperature in this longwall room is about 100° F. Sweat and coal dust have long since turned the faces around me to lumps of coal, in which only the eyes and sometimes the teeth – when there is time for a joke – show a glimmer of light.

The work I have captured in these photographs is hard. Unlike the many people who ruin their bodies through stress or lack of exercise, the miner doesn't have the option of swallowing pills or engaging in leisure sports activities to reconstitute himself. If anything can help, then it is rest. After a long day, sleep often overpowers him immediately. Work in the mine taps the depths of substance.

Miners are paid by shift and ton of coal produced. Each shift is worked by an experienced, finely-tuned group known as a "third," which people above ground would probably refer to as a "team." In a good third, words are hardly needed at all. Enveloped by darkness, dust and heat, the group must work together like a well-oiled machine, automatically, or the coal doesn't move. Nor does anything else, either.

Why do people willingly choose to work under such conditions? Why would someone take a job like this instead of one of those

nice positions sitting in front of a computer? I asked about that. I asked the miners if they would leave the mine if they were offered a different job for the same pay. There were a few who said they would go and many who said they wouldn't. Surely, the miners' responses were influenced by the fact that they know it is unrealistic to expect that they could get a job paying as much with the training they have. But since miners have been going without pay raises for years, while the benefits paid by the Knappschaft, their health insurance, have steadily deteriorated, they haven't been a high-income group for some time.
There is another, decisive factor involved in their choice, and it is understandable only to those who have been down below.

It is no coincidence that miners are referred to as Kumpel. Down here, where they work under difficult, extremely hard and

Translator's note: In the miners' language, the word Kumpel – literally "mate" or "buddy" – denotes the occupation of a pitman or miner and affirms the natural bond that joins one mineworker to another.

very dangerous conditions, every man must be able to rely on his co-worker. Those who fail to look out for others endanger both themselves and their companions in the mine shaft. There is no doubt: Miners' ways are rough. The Kumpel who, looking out for "number one," leaves the others in the lurch or shirks his duties down below is bound to feel the consequences. The worst punishment is being denied a back-scrub –colleagues' refusal to wash a fellow miner's back in the showers. It is tantamount to being branded.

There still exists here something I regard as a "functional social structure." Everyone who works here is a Kumpel or will soon be taught to be one. This does not mean that no one is permitted his own idiosyncrasies or "quirks." Every miner has the right to retain his own individuality, with all of its sharp edges and rough spots, as long as he does his job well and does his part for the others.

Down below, everyone is human.
Another important factor in the miner's choice is the work itself. I have met people who did hard work and were satisfied with themselves at the end of the day. Disregarding the adverse conditions under which they work, it is being able to experience of their own power to perform that counts for them. I have met men with a strong, natural pride in themselves and their work – a pride that has made them free.

Times are bad for the German coal-mining industry. We need only call to mind the large-scale miners' demonstrations held in March of this year – and the egocentric reactions of the politicians, reminiscent of the behavior of feudal lords, astonished by people who were prepared to defend themselves and who had not lost sight of the important things. We need only consider the policies of the Ruhrkohle AG, an organization whose new corporate divisions were built on the achievements of coal-mining but which now prefers to call itself the RAG in an effort to conceal its own origins. Then we recognize how bad things really are in the German coal-mining industry and for the people who work in it. Sometimes it is impossible to avoid the strange feeling that a class of managers is intent upon turning certain social or economic tendencies into a fait accompli of material necessity.

Yet regardless of specific aspects of the debate over the closing of German coal mines, we should not lose sight of two things: where we come from and where we are going.
I was born in the Ruhr region and regard myself as fortunate to have grown up in a part of the country in which tolerance is more than merely a word. Having completed my work on this

theme it is now clearer to me than ever before that coal mining has a great deal to do with the way people are here. People compelled to recognize, day after day, that being considerate of others is in one's own self-interest and that no one can survive without the help of community have learned what it means to be a social being – and not just down below. Our society is still nourished by this awareness today.

Today, we witness the growth in our society of a world of work that does little or nothing to foster a sense of community. Instead, it appears to promote the belief that we can do without it. Increasingly, individuals find themselves virtually alone at isolated, high-tech workplaces. Our offices are scenes of an uncompromising struggle for careers, of mobbing and of the many other manifestations of hard-line egotism. Thus the workplace today is rather a training ground for single-combat thinking and techniques of intrigue. But how can such a context engender a consciousness of community, a will to work for a common interest? For a company? For a nation? To what is the individual expected to develop a sense of belonging under such circumstances? Why should a person recognize a moral obligation (even to pay taxes, for example) towards the community in which he lives if he cannot experience it as such?

The looming threat of the disappearance of work in the mines should make us anxious from a political standpoint as well. We are about to lose another sector of the world of work in which it is not career-orientation but socially-responsible behavior that counts. As we look to the future we should remember that human beings naturally seek a community in which to live and work, a community in which they experience a sense of usefulness and belonging.

Where can we expect to be led by a society whose main purpose is to produce as much as possible with the help of as few people as necessary? Even if the wealth thus produced were to benefit everyone (and not merely an increasingly smaller proportion of the population) – what good is a full stomach to those who feel left out and useless? Should it not be society's goal to keep people both in bread and in work?

No one who looks closely at the pictures of work in the coal mines will fail to wonder whether people should really be required to do this kind of work at all. Yet we should also remember that maximum comfort and convenience do not always produce happiness – neither in work nor in life.

"Glückauf"

Friederike Most

Born in 1965; studied photography at the Fachhochschule Krefeld, earning her diploma in 1995. Friederike Most lives and works in Germany.

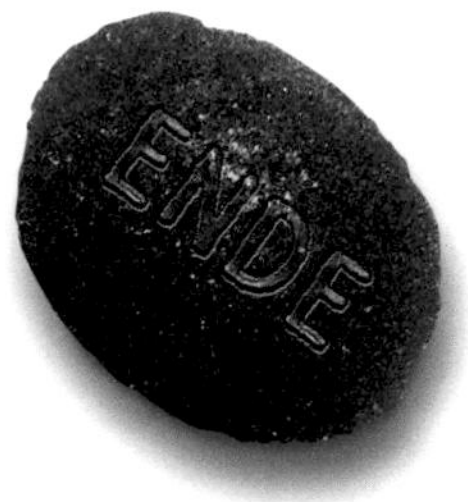

All of the explanations of mining terms used in this book have been cited directly or in paraphrase from the "Kleines Bergbaulexikon," published by Glückauf Verlag, Essen (4th ed., 1983), with the exception of the first lines from a poem by Mr. Döring that appear on page 143. This poem, written in 1926, is printed in Josef Hoffman (ed.), "Der ewige Bergmann," Deutscher Wald Verlag Verlagsanstalt Rheinhausen, 1–5th ed., 1958.

Reproductions and texts copyright by Friederike Most
Translations from the German texts by John S. Southard
Book concept and layout by Friederike Most
Art direction by Guido Widmer
Photolithography and printing by Kündig Druck AG, Baar, Switzerland
Binding by Buchbinderei Burkhardt AG, Mönchaltorf, Switzerland

ISBN 3-908162-89-0